best wishes,
Harold Willis

The Rest of the Story . . .

about Agriculture Today

by Dr. Harold Willis

2nd printing, revised, 1984

Designed and produced by A-R Editions, Inc.,
315 W. Gorham, Madison, Wisconsin

ISBN 0-912311-00-2

Printed in the United States of America

For sale by the author, P.O. Box 692, Wisconsin Dells, Wisconsin 53965

Dedicated to
farmers everywhere,
who work longer
and harder
than most people,
and receive less
than they deserve.

Drawings and photographs
are by the author
unless indicated otherwise.

N P. 113
P P. 117
K P. 120

CONTENTS

The Rest of the Story . . . about Agriculture Today

by Dr. Harold Willis

FOREWORD

Why this Book?

Few people who have heard him can resist Paul Harvey's unique newscasting style—how he probes behind the cut-and-dried syndicated news releases and digs out. . . . "the REST of the story"—the interesting and important details, the ultimate causes and reasons behind the news event.

Very little in this modern 20th century world is what it seems to be on the surface. In almost every facet of this bustling, flashy, and often jarring society, there are important details, causes and reasons that the average person, the one caught up in it all, never hears about.

This is as true of modern agriculture, probably the most important industry in the world (for who can do without food?) as any other part of society. We tend to believe and follow the proclamations of "the experts," those in government, industry, and academia who SHOULD know what's going on and SHOULD advise us to do the things that are for our benefit, for the "common good."

Who else can we trust? Few people can spare the time to spend in libraries, paging through endless books and technical journals, digging out the facts for themselves—to become their own "expert."

Could it be that we don't hear the whole story from the experts? Is their advice for our good? Do they even know the whole story? Most "experts" are specialists, concentrating on one narrow field of "expertise." In other words, they operate with blinders on! They themselves don't have time to become knowledgeable in areas outside their specialty. The number of books and technical articles published every year—no, every month—is truly astounding. No one can read even a small percentage of the material published in his own specialty. So, in many cases, the experts are not even aware of important research and developments in their own area—what their colleagues in Russia or New Zealand are doing, for instance. For this we can hardly blame them.

In other cases, the experts have heard of findings by others, but actually *refuse to accept them*, perhaps because they do not fit their preconceived theories, or because they were not discovered by "professionals" with the approved string of degrees behind their names, or because they are discoveries made years in the past and thus cannot be up-to-date, or for still other reasons. This kind of unprofessional behavior we cannot tolerate.

For these reasons, this book will try to fill you in on some amazing and important details and facts you probably never heard before, to show you reasons and causes—to tell you the REST of the story.

The truth is not always easy to find. That's why we should value it highly once we find it. It isn't lying right out in the open, with a flashing neon sign on it, "TRUTH." It takes digging, sifting, refining, and polishing to get it out. It is hidden and scattered, "here a little, there a little." It has to be put together like a jigsaw puzzle, piece by piece. We are going on a fascinating journey through nature and through man's institutions. Fasten your seatbelt!

For those who don't have time to read all chapters in detail, or who want a quick overview, it is suggested that Chapters 3 through 8 be skipped or saved until last. Numerous boxes scattered through the book provide supplementary information and technical details.

CHAPTER 1

Fiddler on the Roof— Tradition!

To those who have seen the wonderful play or movie, "Fiddler on the Roof," one of the memorable songs is "Tradition!" How often we follow tradition, doing things the way our fathers did or the way our neighbors do, because we are a little bit afraid to break away from the crowd, to try something different. And of course, we don't have confidence in our own knowledge and logic, so we follow the experts' advice—after all, they should know, shouldn't they?

Unfortunately, most school children do not particularly like the study of history, and therefore most of us grow up lacking an appreciation of the past or a desire to investigate it. We are so caught up in the whirl of this scientifically, technologically marvelous 20th century that we think (or are told) that only the most up-to-date methods and products are worthwhile. We have forgotten, or never bothered to find out, how our grandfathers used to farm.

What is considered "traditional" or conventional agriculture is hardly old enough to be dry behind the ears, much less be truly traditional. It is mostly a post-World War II phenomenon. Before we are finished with our journey, we may decide that the *real* traditional agriculture, the way our grandfathers farmed, has more of value to it than the snap, crackle, and pop of "modern" agriculture.

One of the few advantages of growing old is that you can remember what it was like in the "good old days." If you are old enough, or just sit down and think for a moment, or better yet, talk to an "old codger," you will realize that *there are more problems now than there used to be.* Worse insect problems. Worse weed problems. Corn blight and moldy corn. Cattle with mastitis and breeding failure. Even soil is harder, requiring larger tractors to plow it. Most of these troubles have appeared, or at least have become widespread, since World War II. Could there be a connection between this and the fact that modern farming methods have also become widespread—"traditional"—since World War II?

Every problem has a cause—or causes. Every problem also has symptoms, which are signs and results of the cause, but are not the cause. *In order to solve or eliminate a problem, we must attack and eliminate the cause, not the symptoms.* Once the cause is eliminated, the symptoms—and the problem—will disappear. Weeds are not a problem—they are symptoms of an underlying cause. So are insect pests, hard soil, and sick animals. Correct the basic problem and weeds, insect pests, hard soil, and sick animals will disappear, along with a host of other symptoms—nitrate poisoning, dust-bowl conditions, stillborn calves, herbicide carryover, grain drying costs, drought-killed crops, and on and on.

It can be done. It has been done, by a growing number of farmers. It may not happen overnight. It takes time to undo damage and solve problems that have been building up for decades. But improvement can begin right away, and every year the trend should be upward.

Before we can see the causes of our problems and how to solve them, we need to examine how the natural world works, to look at the systems of nature that we work with (*work with*, not dominate) and see how they should operate (if we will only let them).

Also, before we proceed on our journey, let's spell out in more detail what we have called "traditional" or modern agriculture. By this we mean the methods developed or becoming widespread since World War II, based on modern "science" and technology (often war technology). These include (1) the predominant use of inorganic salt fertilizers as the major or sole source of crop nutrients; (2) the use of synthetic chemicals (often very toxic) for weed control (herbicides), pest control (pesticides, insecticides, nematicides, acaricides), animal health (drugs, antibiotics), and increased production (hormones, growth stimulants); (3) certain land use and soil tillage practices (clearing windbreaks, irrigation in dry-climate areas, plowing dry grassland and marginal land, and minimum tillage or no-till methods); and (4) other production methods that interfere with natural cycles and processes (total animal confinement, hybrid varieties, re-feeding manure).

Not all of these methods may be totally wrong under certain circumstances, but if over-used or used under wrong conditions, *they generally boomerang*—they have consequences that are unexpected and undesirable.

STOP

CHAPTER 2

View from the Mountain— Ecology

Before assaulting problems, it is best to back off a distance and be sure we comprehend the total situation—that we see the forest and not just the trees. Then we can better recognize problems, symptoms, causes, and solutions. So the first stop on our journey will be to climb a mountain and take a panoramic view of nature.

Agriculture works with natural systems, with living biological species of plants and animals, and with the natural environment. The branch of biology which deals with all of these things at once is *ecology*, the study of organisms (living plants and animals) and their environment, AND the interrelationships among them.

Nature is incredibly complex. It is composed of millions of functioning parts. Even one cell from any plant or animal is more complex than any man-made machine. Ecology studies organisms and their environment—how they live and function in their environment and how they interrelate with each other and with the environment. We find that these "parts" of nature, like the gears and levers of a machine, have their own functions and that all the parts together form a larger entity, a natural or biological *system*, or to use the terminology of ecology, an *ecosystem*. An ecosystem and its parts operate under certain natural laws. These established laws operate without fail, and if they are broken, a penalty results, just as when a criminal is punished for breaking a man-made law. If we understand more about natural systems and how they operate, we can often eliminate, or better yet, avoid problems (which are the penalty of broken laws).

The food chain. The most important natural system to give us an overview of agricultural problems is the general food chain, the cycle of energy transfer involving stored energy called food. The basic cycle is very simple, and virtually everyone is at least subconsciously aware of it.

Let's start with the organisms we are most familiar with—ourselves, and our domesticated animals. What do we use for food? Most people eat a variety of plant and animal food. What do our domestic animals eat? Plants, almost entirely. What do plants live off of? They make their own food, by the process called photosynthesis, from carbon dioxide, water, small amounts of mineral nutrients from the soil, and light energy trapped from the sun. Where do plants grow? In the soil, from which they obtain water and minerals.

These statements probably aren't much of a surprise. You probably learned them in grade school. We have now built up several steps of our food chain:

These are the most obvious steps and the ones most people, including farmers, are more or less familiar with. But are we finished? Could these parts operate continuously like that? No, we must add several more steps—the REST of the story—in order to complete the picture, like a belt to connect two wheels in a machine. Animals produce what is usually called "waste," mostly manure, plus urine. Most crop plants produce unused stalks, leaves, and roots, which we often call crop residues. These plant and animal "wastes" are not, or should not be called, wastes, for in nature they are recycled. These materials, manure and plant residues, are called *organic matter*. They should be returned to the soil, where they should decompose, or rot, into a substance called *humus*. This decomposed organic matter is a natural plant food, supplying needed nutrients in the forms and at the time plants need them.

The final link in the chain is a multitude of mostly microscopic organisms in the soil which perform the functions of decomposing ("eating") fresh organic matter to form humus and helping to supply nutrients to plants. These soil microorganisms are a vital, but generally overlooked part of the ecosystem. Thus we have built our food chain:

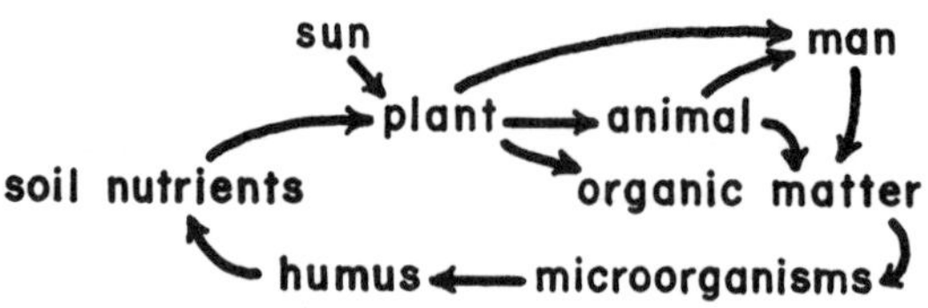

It is a system, a *cycle*. It "runs" on energy just as a man-made machine does, originally energy trapped from the sun by plants, and then passed on to each step in the system. It operates under definite laws. A deficiency in or elimination of one step in the system will decrease or eliminate the next step—and the next—and the next. In other words, the "balance of nature" is upset. And how easily man can throw the system out of balance!

Problems and causes. For example, a deficiency of the mineral element *zinc* in the soil will result in the plants growing in that soil being deficient in zinc.[1] Zinc is a necessary part of certain plant enzymes, called dehydrogenases, which are vital in the metabolic functions of cellular respiration.[2] So a zinc-deficient plant will have impaired cell functions—it will not be a normal, healthy plant. Animals or humans who eat zinc-deficient food will suffer a variety of health problems, depending on conditions. Zinc helps heal wounds and skin irritations. Extreme deficiency can lead to delay in intestinal absorption of carbohydrates and proteins (poor utilization of food), low production of proteins and nucleic acids, skin problems, falling hair, sexual dysfunction, birth defects, and retarded growth. Zinc is a component of insulin, which is necessary for carbohydrate metabolism, and can counteract lead and cadmium poisoning. Zinc is necessary in an enzyme, called carbonic anhydrase, which functions in maintaining equilibrium between carbon dioxide and carbonic acid in tissues, which is important in the blood's carrying of carbon dioxide. Zinc is also necessary in digestive enzymes produced by the pancreas.[3]

What might be the ultimate cause of the zinc deficiency in the soil? There may be plenty of zinc in the mineral materials making up the soil, but it may not be *available* to the plant. For example, at high (alkaline) pH, soil zinc is less available than at low (acid) pH.[4] Also, zinc is not always just absorbed by plants as a simple ion. Zinc, along with several other elements, can be more readily absorbed into plants in *chelated form*, an inorganic ion held and carried by a larger organic molecule.[5] Natural chelators include certain organic acids, humic acids, found in humus.[6] Microorganisms are necessary to produce adequate humus in soil.

[1]Gilbert, F. A. 1948. *Mineral Nutrition of Plants and Animals*. U. of Oklahoma Press, Norman, OK.

[2]Goss, J. A. 1973, *Physiology of Plants and their Cells*. Pergamon Press, NY.

[3]Bogert, L. J., G. M. Briggs & D. H. Calloway. 1966. *Nutrition and Physical Fitness*, 8th ed. W. B. Saunders Co., Philadelphia.

Pike R. L. & M. L. Brown. 1975. *Nutrition: An Integrated Approach*, 2nd ed. John Wiley & Sons, NY.

Rodale, R. 1976. *Zinc—How Much is Not Enough*? Prevention. Sept., p. 23–28.

[4]Brady, N.C. 1974, *The Nature and Properties of Soils*, 8th ed. Macmillan Publ. Co., NY.

[5]Noggle, G. R. & G. J. Fritz. 1976. *Introductory Plant Physiology*. Prentice-Hall, Englewood Cliffs, NJ.

[6]Albrecht, W. A. 1962. *The Healthy Hunzas, A Climax Human Crop*, p. 426–431 *in* The Albrecht Papers, 1975, C. Walters, Jr., ed.

So what is the *cause* of sick animals or people in this hypothetical example? Is it the zinc-deficient food they ate? Or the low zinc availability in the soil the plants grew in? No, these are just symptoms. The real cause is something wrong in the soil. Perhaps toxic agricultural chemicals killed the microorganisms which produce humus. Or perhaps wrong fertilizers have altered the pH or "burned up" the humus. Or perhaps not enough manure and plant residues were returned to the soil. Whatever the cause or combination of causes, they can be eliminated or remedied, which will lead to adequate zinc availability, which will lead to healthy plants, which will lead to healthy animals and people. *How simple, once we understand the biological system.* Notice also, how we assembled the parts of the story from a variety of sources, here a little, there a little.

A concise slogan used by ecologists to explain what their science is all about is, "Everything is connected to everything else." This beautifully describes how natural systems are organized and function. A change in one part of the system affects the whole system to some degree.

Now that we have climbed our mountain of ecology and taken a quick look around at the landscape, let's go down the mountainside and plunge into the forest and examine the trees. Let's journey through the food chain and look at the parts and processes in more detail.

First stop—basement!

CHAPTER 3

The Foundation— Soil

My, it's dark down here in the soil. No wonder most people know so little about it. But that's why we're here, so let's learn.

Soil is the absolute basis of agriculture, and thus of all human existence, for as we have seen, we either eat plants grown in soil or animals which eat plants grown in soil. Our soil has been called our most important national resource. Wise use and management of the relatively thin upper layer, the topsoil, is vital for maintaining good health and a high standard of living.

But through misuse, about 7–10 *tons* of topsoil per acre are being lost to erosion each year in the midwest (the figure can be much higher in the worst areas). It may take several hundred years for one inch of soil to

form. Obviously, we can't keep on sending our topsoil down the river much longer.

Besides that, most once-fertile soil is now polluted by toxic substances, along with the groundwater and our wells. We are literally fouling our own nest, destroying the hand (or land) that feeds us! We had better understand more about this important, but neglected and abused part of the food chain. Let's take a tour through this dark and mysterious land that lies beneath our feet.

Parts. What is soil? It is a very complex substance, not "just dirt." Soil is a mixture of several components, sometimes defined as that part of the earth's surface capable of supporting plant life. It originally formed by the weathering of the rocks of the earth's crust. Typical soil contains approximately the following proportions of four constituents (H.D. Foth & L.M. Turk, *Fundamentals of Soil Science*, 5th ed., 1972):

1. *Minerals* (about 45%), some of which are insoluble and not used by plants (sand, clay, iron oxides) and others which are soluble and provide valuable plant nutrients (calcium, potassium, magnesium). Mineral particles range in size from coarser gravel and sand to finer silt and the smallest clay particles, which are in the size range called colloidal.

2. *Water* (about 25%), which is needed as a part of plant cells and to dissolve and carry nutrients. Too much water in soil can exclude needed air.

3. *Air* (about 25%), which provides oxygen to roots and soil microorganisms, and nitrogen to nitrogen-fixing bacteria. Good aeration is vital to fertile soil.

4. *Organic matter* (about 1-5%), which includes the living soil organisms and the dead organic matter which decomposes to form humus. Humus has been broken down to very small particles in the colloidal size range. A "good" soil should have 2-5% organic matter, and up to 10% can be beneficial.

Colloids. In the above four components of soil, two contain colloidal particles, clay and humus. Colloidal particles, which are particles less than 0.002 mm (1/5000 inch) in size, are important in soil because they have a great ability to hold certain plant nutrients. Humus colloids can hold three times the nutrients that clay can. This is one of the reasons humus is so valuable in soil.

Tilth. A "good" soil should have a loose, almost spongy texture because the tiny soil particles (sand, silt, clay) are clustered into small "crumbs." This condition is called good soil structure, or good tilth. Soil structure affects the ease of water penetration and aeration, root growth, the activity of soil organisms, and the availability of nutrients. Good tilth is generally only found in the upper layers of soil, with the

Roots have trouble penetrating a hardpan.

lower, harder layer often being called a "hardpan," "plowpan," or "claypan." Soil of good tilth is easy to plow, soaks up water like a sponge, and resists erosion.

Factors that contribute to good soil structure include, freezing and thawing, wetting and drying, penetration by plant roots, animal burrows, soil colloids, and most important, a "glue" secreted by roots and soil microorganisms. This is one of the many reasons soil organisms are important.

In poor soils, too much tillage can destroy good tilth, as can leaving the soil without a vegetative cover to cushion raindrops and to encourage microorganism growth. Also, too acid or too alkaline conditions can destroy good tilth.

pH. One soil condition that agricultural "experts" concern themselves about excessively is pH. This is a measure of acidity or alkalinity of any substance using a scale of numbers, from 0 (most acid) to 14 (most alkaline), with 7 being neutral. The pH of some common materials includes: lemon juice = 2, vinegar = 2.5, black coffee = 5, pure water = 7, sodium bicarbonate solution = 8.2, ammonia water = 11. Soil pH generally varies from 4 to 10, but most crops do best under slightly acid conditions (6.0-6.8). Soil pH affects the availability of nutrients, which may be connected with nutrient deficiencies and toxicities (for example, manganese can be toxic to plants at pH 4.5 or below). Soil pH also affects the types of soil organisms and their ability to flourish, including those that fix nitrogen. Most bacteria cannot live in very acid conditions, while many fungi can.

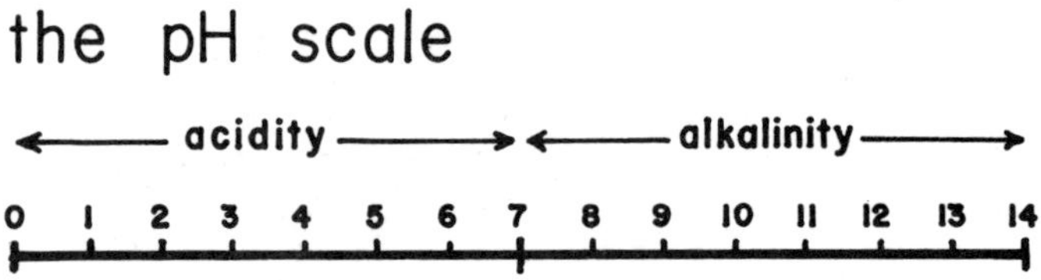

The traditional belief is that soil acidity is bad and should be counteracted by applying lime. But some acidity is necessary for plants to absorb certain nutrients from soil colloids. The experts usually base their liming recommendations on a single soil pH test, but testing several times during a year will reveal that pH can change markedly through a growing season, even daily. Also, various fertilizers and soil conditioners can have short-term and long-term effects on pH. So, as long as the pH does not become extremely acid or alkaline, pH "correction" and testing are not as important as some seem to think. The use of lime to "sweeten" soil is a case of doing the right thing for the wrong reason. Crops benefit more from the calcium supplied than from pH control. Care must be taken to only use pure calcite lime ($CaCO_3$) rather than dolomite ($CaCO_3 \cdot MgCO_3$), often called ag-lime, because the magnesium in dolomite depletes available nitrogen, pound for pound (see Chapter 11).

Water. The movement of water in soils is little known or often misunderstood by most people. The average person assumes that water simply moves downward in soil, but if water is applied to soil at a single point, it defies the law of the gravity and moves just as fast horizontally as vertically; thus it soaks into the soil in a spherical pattern. Of course, rain normally falls all over the surface, not at a single point, so generally a horizontal "front" of water will soak in.

One of the most amazing facts of water penetration is the barrier effect of a soil of different texture, such as a subsoil "hardpan" or simply a change in texture, such as from fine to coarse. When water encounters such a change in texture, its rate of penetration will be considerably slowed, and in reverse, later rise of water to supply roots will also be slowed by such a barrier.

The role of porous organic matter, such as plant residue, in absorbing water can be important under the right conditions. Organic matter which is thoroughly incorporated into the upper layers of soil or which is

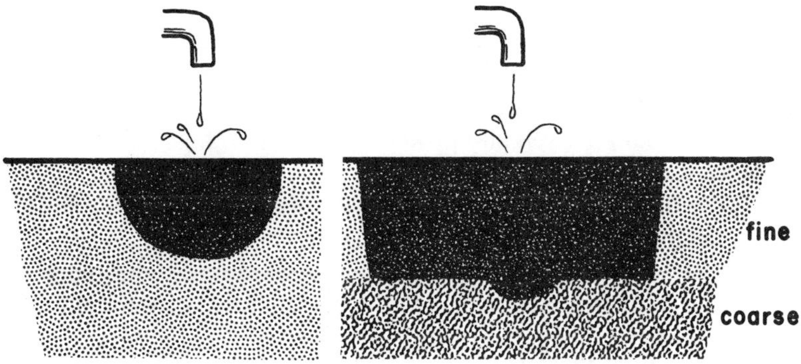

Left: water from a point source penetrates uniform textured soil in all directions; Right: a layer of coarser soil forms a barrier and retards water penetration.

partly exposed to the surface acts as a wick to increase water penetration. A compact plowed-under mass of organic matter acts as a barrier.

Organic matter. The importance of organic matter in soil cannot be overemphasized. William A. Albrecht, former head of the Soils Department at the University of Missouri, once called organic matter the "constitution" of the soil. As mentioned previously, soil organic matter consists of the dead decomposed humus and the living soil organisms. First let's look at humus, which has been called "the most important source of human wealth on this planet" (S. Waksman, *Humus*, 1938, p. 414).

Humus is a structureless colloidal material resulting from the decomposition (humification) of any type of dead organic matter (mostly plant residues and manures). It is a complex chemical mixture including proteins, lignin (originally part of plant cell walls), fats, carbohydrates, and organic acids. Its great ability to hold nutrients because of its colloidal nature has already been mentioned. The beneficial aspects of humus include:

1. It provides a storehouse of essential plant nutrients; for example, it stores over 95% of the nitrogen, 60% of the phosphorus, and 98% of the sulfur available to plants.

2. It helps make some nutrients more soluble and available to plants. Because of increased microorganismic activity during the warmer months, nutrients are released at the time of the plants' greatest need. Acids in humus also slowly dissolve soil minerals and release nutrients.

3. It contains substances that stimulate plant growth and improve crop quality and resistance to pests and diseases.

4. It provides a high water absorption and holding capacity because of its spongy nature.

5. It contributes to good soil structure (good tilth) by helping make soil crumbly and porous. It also reduces wind and water erosion and makes soil easier to work.

6. It buffers the soil and protects plants from high salt levels, toxic chemicals, and drastic changes in pH.

7. It provides food for the beneficial soil organisms, especially in the "pre-humus" form of fresh organic matter.

So we see that humus is a marvelous substance, with many functions, and "worth its weight in gold." Now let's look at its equally important partner, the living part of the soil.

Fertile soil is literally teeming with an amazing variety of plant and animal life, most of which are microscopic and thus little known nor appreciated by the average person. They form a valuable "work force," performing a multitude of chemical transformations, as well as many other services. An idea of their abundance can be gained from the fol-

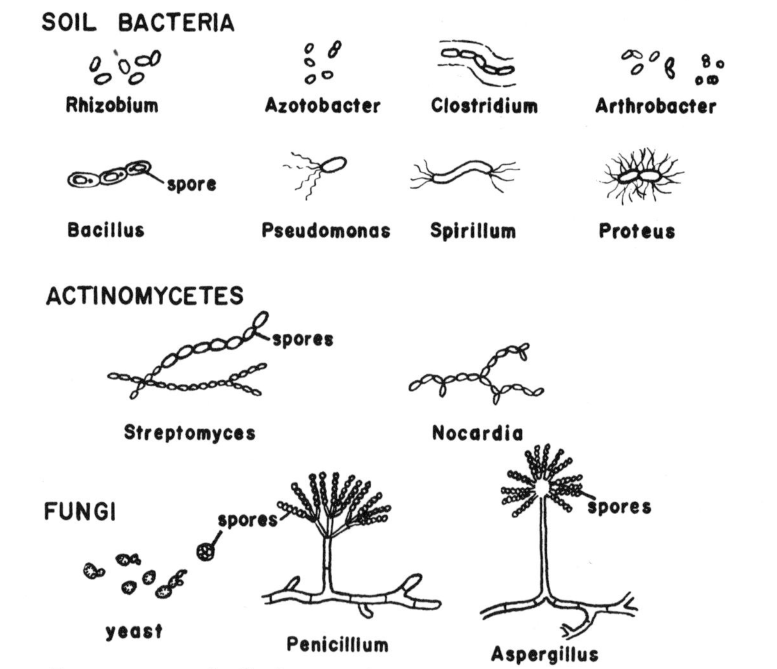

Common types of soil microorganisms.

Nitrogen Fixing

The ability to transform the chemically stable gas nitrogen, into organic nitrogen compounds is found in only a few groups of microorganisms: about 20 types of bacteria, 10 types of blue-green algae, two types of yeasts, and probably one type of actinomycete. The most important to agriculture are the bacteria, which include both aerobic and anaerobic species, most of which are common in the environment and live in soil independently of plants. The best known of these free-living bacteria are the aerobic *Azotobacter* and the anaerobic *Clostridium*. These bacteria grow best when there are carbohydrates, which can come from organic matter (manures and crop residues), added to the soil as a source of food. "When large quantities of carboyhydrates are incorporated into soil, significant increases in combined nitrogen occur" (T. D. Brock, *Biology of Microorganisms*, 1970, p. 578). In laboratory test jars of composting dairy cow manure, I have found an increase in nitrate nitrogen of over 60 times due to the action of nitrogen-fixing bacteria.

The best known nitrogen fixing microorganism is *Rhizobium*, the type of symbiotic bacteria which grows in the root nodules of legumes, such as alfalfa, clover, beans, and peas. Neither the legumes nor the bacteria alone can fix nitrogen, only when they "cooperate" can this feat be done. The formation of root nodules is amazing. The *Rhizobium* bacteria are among the common rhizosphere organisms that grow on plant roots. One of the materials secreted by legume roots is the amino acid tryptophan. *Rhizobium* changes tryptophan into the plant hormone indoleacetic acid, which causes the root hairs to curl. Then some of the *Rhizobium* change into highly mobile "swarmer cells" which work their way through the root hair's cell wall. Once inside, the swarmer cells grow into a thread-like structure that invades cells of the root's cortex region. If a root cell happens to have twice its normal number of chromosomes (a small number of such cells are always present), it is stimulated to begin dividing and producing a nodule. The bacteria then rapidly multiply and fix nitrogen inside the inner nodule cells. A "healthy" nodule will have a pink or red color inside because of a red protein called leghemoglobin, which is similar to the red hemoglobin that carries oxygen in our blood.

The *Rhizobium* bacteria can survive in the soil for many years, waiting to reinfect new legumes, IF soil conditions are favorable. Needless to say, anything farmers add to their soil which kills soil life cheats them out of about 80 pounds per acre of free nitrogen fertilizer.

Several types of plants other than legumes can have nitrogen fixing nodules that contain other microbes than *Rhizobium* (blue-green algae and a probable actinomycete). Research is currently being done to try to get non-legume crops to grow nodules.

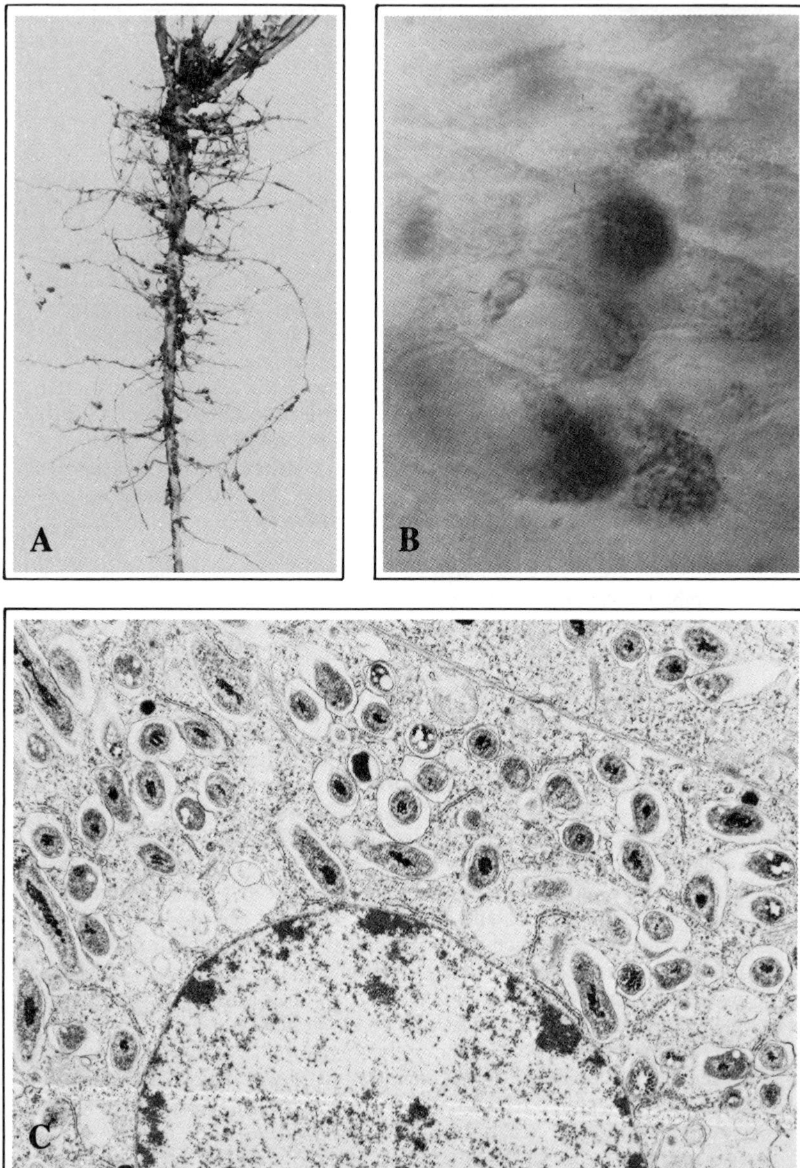

A. Root nodules of alfalfa; B. nitrogen-fixing bacteria inside alfalfa root nodule cells, magnified 100 times under light microscope; C. nitrogen-fixing bacteria (smaller round and sausage-shaped objects) inside a nodule cell (large rounded structure at bottom is the nodule cell's nucleus); magnified 5000 times under electron microscope *(photo courtesy Dr. Shiv Tandon).*

lowing table, which shows figures for a temperate grassland soil (from L. M. Thompson & F. Troeh, *Soils and Soil Fertility*, 4th ed., 1978):

organism	*number/acre*	*lbs./acre*
bacteria	800,000,000,000,000,000	2600
actinomycetes	20,000,000,000,000,000	1300
fungi	200,000,000,000,000	2600
algae	4,000,000,000	90
protozoa	2,000,000,000,000	90
nematodes	80,000,000	45
earthworms	40,000	445
insects & other arthropods	8,160,000	830

The first four types of organisms are plants and the last four are animals. The most important are the bacteria, actinomycetes, and fungi. Bacteria are single-celled organisms that can live in either aerobic (aerated, with oxygen) or anaerobic (without oxygen) conditions, but generally cannot live in very acid soils. The aerobic bacteria are the beneficial kind in general. Some of the beneficial activities of soil bacteria include helping decompose organic matter to form humus, converting inorganic chemicals into useful plant nutrients, detoxifying (breaking down) man-made toxic chemicals (herbicides, pesticides), and fixing (trapping) nitrogen from the air for later plant use.

Actinomycetes are filamentous (thread-like) organisms that are considered to be intermediate between bacteria and fungi. They require aerobic, neutral or slightly alkaline soils. They aid in producing humus.

Fungi, or molds, are filamentous, mainly aerobic, and can tolerate fairly acid conditions. Besides those helping to produce humus, one particular group of fungi, the mycorrhizae, are especially important to plants. They are symbiotic and live partly inside the roots of many kinds of plants. They help roots absorb nutrients (especially phosphorus and nitrogen), secrete growth promoting hormones, and help protect roots from disease organisms.

Other important functions of soil organisms besides those already mentioned (decomposing organic matter and man-made toxins, fixing nitrogen, making nutrients available to plants, producing growth hormones, and fighting plant diseases), include temporarily "tying up" nutrients from minerals and organic matter in their own bodies, as part of the proteins, carbohydrates, and fats of their cells. By doing so, they counteract the loss of some nutrients from the soil by leaching into the groundwater. When the organisms die, the nutrients will be made available to the plants, which are thus fed slowly over the growing season. Another important function of soil organisms is, along with humus, pro-

Mycorrhizae

The word mycorrhiza literally means "root-fungus." Mycorrhizae are special symbiotic fungi that live on or in the roots of plants. At first, scientists thought that they were a disease and that they were found on only a few kinds of plants, but now it is known that nearly all plants can have mycorrhizae and that they are a very beneficial kind of fungus, not a disease.

There are two kinds of mycorrhizal fungi. The ectomycorrhizae grow around the outside of the plant's roots and between root cells, but not inside them. They are mainly found on trees. Roots that are "infected" with ectomycorrhizae have a "fatter" appearance than non-infected roots. Some of the species of mushrooms which always grow around certain trees, both poisonous and edible ones (including the gourmet's delight, truffles), are actually the fruiting bodies of ectomycorrhizae. The other kind, called endomycorrhizae, occur on plants other than trees, and actually grow inside some of the cells of the plant's roots (*Crops & Soils Magazine*, Aug.-Sept. 1975, p. 15). *continued. . .*

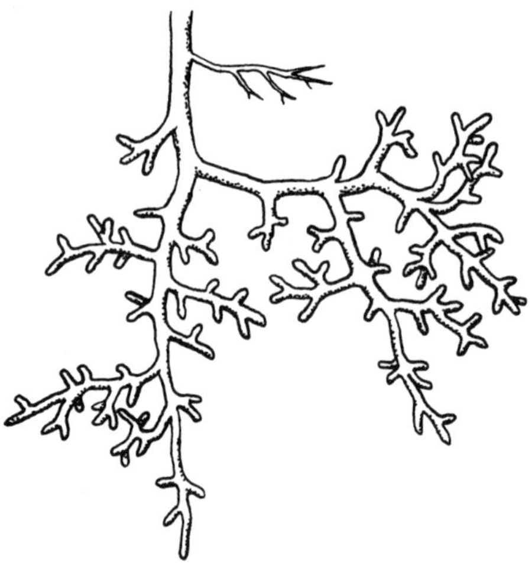

Tree roots with mycorrhizae have a stumpy appearance, while non-mycorrhizal roots are slender (top center).

Mycorrhizae, continued

Mycorrhizae perform a number of valuable services for the plant with which they are associated.

1. They increase the plant's absorption of nutrients, especially phosphorus and nitrogen, and also immobile elements such as zinc, copper, and molybdenum, and the ammonium ion. In a 1937 study of white pine, mycorrhiza-infected trees absorbed 1.24 mg nitrogen, 0.196 mg phosphorus, and 0.744 mg potassium (% of dry weight), compared to non-mycorrhizal tree figures of 0.85 mg nitrogen, 0.074 mg phosphorus, and 0.425 mg potassium (J.L. Harley, *Mycorrhiza*, 1971). Nutrient absorption can be increased partly by the fact that the fungal threads increase the absorptive area of the root and can reach out farther than the plant's root hairs. Also, mycorrhizae aid in nutrient absorption by utilizing nutrients tied up in soil organic matter. Thus mycorrhizae can change some nutrients into forms useable by plants.

2. Mycorrhizae produce hormones that stimulate the growth of their host plant (V. Slankis, *Annual Review of Phytopathology*, vol. 12, 1974, p. 448). This plus the increased nutrient absorption just mentioned, can combine to increase plant growth significantly—sometimes by several hundred percent.

3. Mycorrhizae can increase water absorption by a plant because they can grow at lower soil water levels than plants (J.L. Ruehle & D.H. Marx, *Science*, Oct. 26, 1979, p. 421).

4. Mycorrhizae can protect roots from diseases, either by forming a protective barrier or by producing antibiotics.

In return for its services, the fungus receives a supply of carbohydrates and certain other nutrients from its host plant; thus both partners in the symbiotic relationship benefit. Mycorrhizae benefit plants most when the plants are growing in poor soil, particularly phosphorus-deficient soil. Plants growing in fertile soil do as well with or without mycorrhizae (J.L. Harley, *Mycorrhiza,* 1971).

Even though nearly all crops *can* have these helpful mycorrhizae (cabbage, spinach, and garden beets are exceptions), many farmers are missing out on their valuable services because their soil is not hospitable for their existance. Mycorrhizae, as well as other beneficial soil microorganisms, need the following conditions:

1. adequate moisture
2. oxygen
3. proper temperature
4. proper pH, not too acid or too alkaline
5. adequate soil organic matter
6. freedom from fungicides or other soil sterilants
7. a low concentration of salts, including soluble fertilizer salts (especially nitrates) (*Crops & Soils Magazine*, Aug.-Sept. 1975, p. 16; V. Slankis, *Annual Review of Phytopathology*, Vol. 12, 1974, p. 437-448).

Chemical Warfare in the Soil

Few people, even few agronomists and soil scientists, are aware of the complex interactions among the many species of microorganisms inhabiting soil, and the interactions between them and plant roots.

Besides the dozens of species of bacteria, actinomycetes, and fungi that live on the soil particles away from plant roots, there are other species which only live in the film of sticky material secreted by plant roots (root exudate) (*Science*, 1982, vol. 216, p. 1377). The region close to roots is called the rhizosphere, and the rhizosphere microorganisms play an important role in the well-being of the plant [the special mycorrhiza fungi and the nematode-trapping fungi are covered in detail in other boxes, in this chapter and in Chapter 16].

Soil conditions are constantly changing, both on a large scale (temperature, moisture, pH, etc.) and on a microscopic scale. And different conditions may exist a few feet or a fraction of an inch away. For example, if the pH is above 6, the nitrogen-fixing bacteria *Azotobacter* can flourish. Well aerated soil will allow aerobic bacteria to grow, as well as fungi and actinomycetes. But a heavy rain which accumulates in a low area will exclude air and allow harmful anaerobic denitrifying bacteria, such as *Pseudomonas*, to release nitrogen, while fungi, actinomycetes, and aerobic bacteria become temporarily dormant.

Besides the nitrogen taken from the air by nitrogen-fixing bacteria and algae, other soil microorganisms aid plants by helping to make soil nutrients available (partly by producing acids that "break out" mineral nutrients) and by producing various substances useful to plants—amino acids, growth hormones, and vitamins.

But the most fascinating way in which soil microorganisms interact with plants and other microorganisms is by various chemicals that they produce and secrete by metabolic processes—a sort of chemical warfare. Certain common soil bacteria, actinomycetes, and fungi can produce substances that inhibit or kill other species of microorganisms, especially those that cause plant diseases. Sometimes they only do this when the disease-

continued. . .

moting good soil structure. The "glue" that soil microorganisms produce, which is a type of sticky carbohydrate, has been previously mentioned. It holds soil particles together, forming "good crumb structure;" loose, crumbly, porous soil.

The humus and soil life work together as a team and perform all these valuable services—and the best part is, they work for free, if only given half a chance! Why would anyone want to dump toxic materials on

Chemical Warfare in the Soil, continued

causing species is present. Such human-used antibiotics as penicillin, streptomycin, and tetracycline are grown from the soil fungus *Penicillium* and the actinomycete *Streptomyces*. The Russian researcher, N.A. Krasil'nikov, in his book, *Soil Microorganisms and Higher Plants* (1958, English edition 1961), states that there are natural microbial enemies of all species of soil microorganisms and that they occur by the hundreds of thousands or millions *per gram of soil*! (This is assuming that the soil is "alive," and not sterilized by wrong agricultural practices.) He states, "Through use of their metabolic products antagonistic microbes suppress their competitors, removing them from the substrate and thus exerting a definite selective action. To a certain degree, microbial antagonists regulate the formation of microbial coenoses (colonies) in the soil in general. They play an important role in the improvement of soils, in the so-called process of self-purification of soils. The removal of harmful pathogenic and phytopathogenic flora and fauna is accomplished by microbial antagonists. . . . Antagonists, therefore, can be considered as one of the powerful factors governing soil fertility and plant-crop abundance." (p. 364-365).

It has been found that the contents of plant cells (the "sap") have the ability to suppress the growth of bacteria and fungi (see box, Plant "Antibodies," Chapter 16). N. A. Krasil'nikov believes that organic substances produced by soil microorganisms, when absorbed through the plant's roots, increase this natural immunity of plants to infections. He finds that the most practical way to insure that the soil has these beneficial organisms is for it to be rich in humus and by fertilizing with manure or compost (p. 370-371, 384-385). More recent research has zeroed in on natural iron chelates produced by rhizosphere bacteria. The theory is that these chelates tie up much of the iron, making it unavailable to disease-causing microbes—sort of starving them out (*Science*, 1982, vol. 216, p. 1379). Methods for large-scale application of growth-promoting and disease-suppressing microbes are being developed; seed inoculation has been the most successful so far. Yield increases of up to 30 to 144% have been reported.

his land that will kill or inhibit this volunteer army? Are the experts warning us about this?

For the next stop on our journey, let's take the elevator one floor up and wander about in the green jungle of the plant world.

CHAPTER 4

Food Factory— Plants

Here we are at ground level, walking around in the wonderful sunlight. Makes you feel like stretching your arms out and soaking up some sun. That's it, go ahead, you're already starting to think like a plant!

Plants are such different organisms from ourselves and the animals that we are more familiar with, that they seem to live in a different world. They have no eyes or ears, no bones, nerves or muscles. They can't walk or run around from place to place. But believe it or not, plants are sensitive to light and vibrations, have a supporting skeleton, exhibit electrical activities, and do considerable moving. They are living, "breathing," growing organisms, and anyone who makes a considerable part of his living growing them should know more about them.

To understand more about how plants function, let's shrink ourselves down to the size of a water molecule (only about 0.000006 inch) and take a trip through a plant. A living plant is very much like a bustling city, only much more complex. It is built up of thousands of units called *cells*, as a city is largely composed of buildings—some are factories, some warehouses, some power plants, some houses. Similarly, different cells have different functions: some are "factories," some for storage, some forming a transportation system, like the plumbing of a city. The people and workers in the city would correspond to the many kinds of active ions and molecules, enzymes and energy carriers, that help "run" a living plant. A plant is incredibly complex, so we will only hit the highlights, especially those that relate to agriculture.

To begin our trip through a living plant, we will have to once more venture down into that murky realm of the soil, which we recently visited, but this time we will take a different perspective. If we were the size of a water molecule (0.000006 inch), we would find the soil to be an awesome and bizarre place indeed. We would probably be oozing along in a film of water on the surface of a huge, platelike colloidal clay particle, a

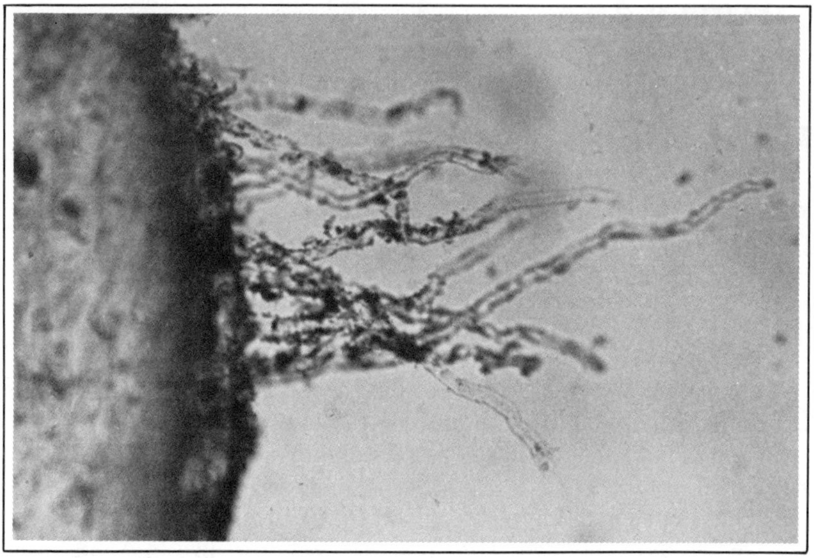

Root hairs of corn, magnified 100 times under the microscope.

gigantic 0.005 inch across. Everything would be in motion, for atoms, ions, and molecules of all matter are constantly vibrating and floating around. We would see fuzzy-looking spheres and lumps of various sizes floating and zooming by. These would be ions and molecules of water, hydrogen, carbonate, calcium, potassium, ammonium, nitrate, and so on. [If you are not familiar with these chemical terms, consult the Glossary and the Appendix.]

We would also have to duck out of the way to avoid being devoured by immense (0.00002 to 0.004 inch) monsters snuffling and gurgling about, such as bacteria and protozoans (one-celled animals), or even an incredibly gigantic, snakelike nematode, or threadworm, 0.04 inch long.

Such is the world of the rhizosphere, the zone close to a plant's roots. Plants primarily absorb two things from the soil—water and nutrients. If a plant doesn't have enough of either, the farmer may as well start over, because he won't get a crop. So we should know a little about how these substances are taken up by plants.

Water. The entrance of water into plants is controlled by osmosis. Osmosis involves the relative percent of water on the outside compared to the inside of a cell. Water moves (diffuses) from wherever it is more concentrated (higher %) to where it is less concentrated (lower %). Thus if there is more water in the rhizosphere than inside a root, it will easily and automatically "seep" (diffuse) inside. Plants have a marvelous set of tiny root hairs at the tip of each root to add surface area for

> In 1937, H. J. Dittmer measured the roots of a single rye plant. He found that the main and smaller roots had a total length of over two million feet (387 miles) and had a surface area of 2,554 square feet, while the tiny root hairs attached to the smaller roots had a total length of 6,604 feet (1.25 miles), but a surface area of 4,321 square feet.

better absorption. The only complicating factor is that some water molecules are so tightly held on the surface of colloidal soil particles that they can never be absorbed by roots.

Every farmer and gardner has experienced the fact that when plants do not absorb enough water, they wilt and die. But is this always caused by drought—not enough water in soil? That can be one cause, but the same result can happen when there are too many salts dissolved in the soil water, lowering the *relative percentage* of water outside the root cells. When there is a higher percentage of water inside the cells than outside, reverse osmosis will occur, the roots will lose water rather than absorb it, and root damage or death of the plant will result. Salt damage is not just a problem in the alkaline and saline soils of the western U.S.; it can happen anywhere if a farmer applies too much salt fertilizer or manure to the soil (manure contains salt from the animal's diet) [see Chapter 12 for more details]. Plenty of rain can alleviate the situation by diluting or leaching excess salts.

One other subject relating to water use by plants is humus. Remember one of the functions of humus listed in Chapter 3 was absorbing and holding water, like a sponge. A soil with plenty of humus will produce beautiful crops even in a drought year.

Back to our trip through the plant. Once water reaches the central "core" of a root, where the "plumbing" is (the vascular tissue, xylem and phloem), it enters into the water-carrying tissue, the xylem, for a quick elevator ride up the roots and stem to wherever water is needed in the plant.

Nutrients. But wait a minute. The roots absorb two materials—water AND nutrients. We next must learn something about nutrient uptake. Here again we get into the complications of soil chemistry. Plants need a variety of chemical nutrients. The essential plant nutrients can be divided into the macronutrients, needed in larger amounts, which are nitrogen, calcium, phosphorus, potassium, magnesium, and sulfur: and the micronutrients, or trace elements, *just as important*, but needed in very small amounts. The micronutrients are iron, copper, zinc, cobalt, boron, manganese, molybdenum, and chlorine.

These nutrients can be taken into a plant in various forms, some as simple ions (atoms with a positive or negative electrical charge, such as potassium, calcium and zinc), some as molecular ions (the nutrient atom combined with others, such as nitrate, ammonium, and sulfate), as chelates (the nutrient atom held by a larger organic molecule, such as iron, zinc, copper, manganese, and cobalt chelates), or as organic molecules that are precursors of proteins or other cell components (such as phenol, amino acids, amines and indole). "Traditional" agriculture as commonly taught in schools today virtually ignores the latter two, and as-

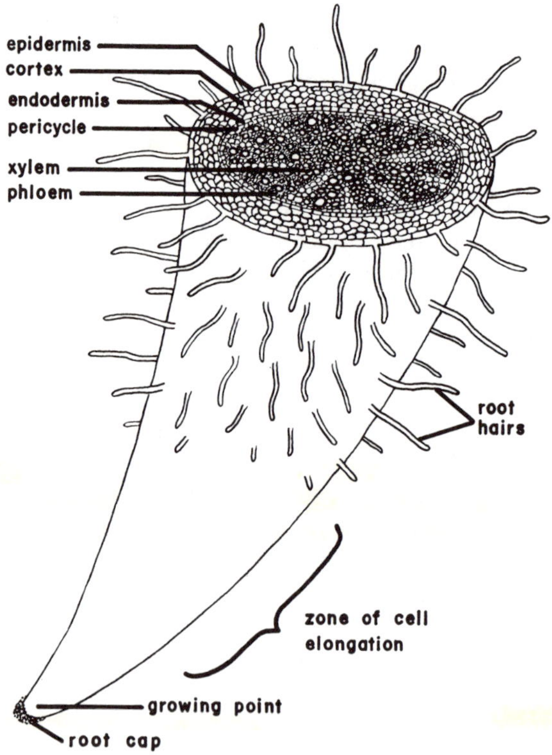

A root is composed of several parts. The root cap protects the growing point. The many root hairs add absorptive surface area. The epidermis covers the exterior. Internally, the cortex stores food, the endodermis regulates inward passage of absorbed materials, the pericycle produces branch roots and additional xylem and phloem, the xylem carries water and minerals upward, and the phloem carries food downward.

sumes that a plant can absorb only inorganic ions and still produce quality crops.

But back to our story of nutrient uptake. The inorganic ions that carry a positive charge are involved in an interesting phenomenon called base exchange, or sometimes cation exchange (a positively charged ion is called a cation, a negatively charged ion is called an anion). The clay and humus colloidal particles carry a negative charge on their surface, which attracts and loosely holds positively charged ions, some of which are plant nutrients. Since they are loosely held, they can easily trade places with another positive ion in the vicinity. This is base exchange. One of the by-products of the metabolism of plant roots and soil microorganisms is hydrogen ions, which have a positive charge, and which can be exchanged for nutrient ions. The more active a plant is, the more nutrients it needs and the more hydrogen ions its roots produce, and the more nutrients it can absorb by base exchange. Isn't that a beautiful,

Chelates

The word *chelate* comes from the Greek word *chela*, or claw, referring to the holding ability of these chemicals. Chelates are organic molecules that are able to hold and release certain metal ions, including such plant nutrient elements as calcium, iron, magnesium, cobalt, copper, zinc, and manganese. These elements are more easily absorbed by plant roots in chelated form than not chelated (and animals can absorb them more readily through their intestinal walls).

Natural chelates are produced by soil microorganisms and are abundant in humus; they include various organic acids. The agricultural scientists, trying to improve on nature, have developed a host of synthetic chelates, the most commonly used being EDTA (ethylene diamine tetraacetic acid). These are sometimes promoted as "organic" chelates, which they are according to the chemical definition of organic—but they're not natural, and evidence is accumulating that they can be ineffective and have harmful side effects. An article by Brad Hinckley in the September 1981 issue of *Acres U.S.A.* (p. 32-33), points out that since the synthetic chelates are alien molecules, plants can only absorb them slowly. Also, after the chelating molecule releases its "payload," it may "latch onto" other nutrients in the plant, thus making them unavailable. For example, synthetic iron chelates cause a manganese deficiency and lower zinc and copper levels; EDTA "grabs" calcium ions and thus upsets the calcium-potassium balance. Synthetic chelates also put undesirable elements into the soil, namely sodium and chlorine, which are used in their manufacture. Finally, synthetic chelates are only stable under a very narrow pH range.

All of which again only goes to show, "Nature's way is best."

self-regulating system for supplying just the right amount of nutrients at the right time? If there's a way to foul up something, someone will do it. Too heavy an application of the highly soluble salt fertilizers will "flood" the soil water with so many ions that the plant will be forced to take in more than it needs. This can upset the plant's metabolism and in some cases be toxic to the animals that eat it (an excellent example being nitrate poisoning from animal feed grown with too much nitrogen fertilizer). We also see from base exchange another reason for the value of soil organisms—releasing hydrogen ions.

In general, the plant nutrients are carried into roots along with the flow of water that we have seen earlier, but independently of the laws of physics controlling water uptake, controlled instead by laws governing colloids and ions. The special mycorrhizal fungi that were covered in Chapter 3 may help in absorbing phosphorus and nitrogen if they are present. Plant roots have an interesting layer of cells just outside of their "plumbing" region, called the endodermis, which acts something like a filter, regulating the passage of water and nutrients from the outside to the "plumbing" inside. The endodermis may help protect the plant against some toxic elements or help the plant concentrate needed nutrients and reject abundant but unneeded ones.

Foliar uptake. Few people realize that plants not only take up water and nutrients through their roots, but also through their above-ground parts, especially their leaves. Plants can take minerals from the air in the form of dust particles and floating ions. G. Ingham of Pretoria, South Africa, believes the air contains all the nutrients needed by plants; a year's rain can deliver up to 30 pounds of nitrogen, 22 pounds of phosphorus, 338 pounds of calcium, and 41 pounds of sulfur per acre (*How to Grow Vegetables and Fruits by the Organic Method*, 1975, p. 107–109). These can be absorbed directly through the waxy outer cuticle which covers the outer cells or through the tiny openings on the surfaces of leaves, called stomata or stomates. Most plants have many tiny hairs on their leaves. It has been found by Dr. Philip S. Callahan (see his book, *Tuning in to Nature*, Devin-Adair Publ. Co., 1976) that plant hairs have an electrical charge because of their outer waxy covering, which acts as an electret (a material with a permanent electrical charge). He has found that similar wax-covered hairs of insects attract molecules out of the air. Could these plant hairs also act as tiny "antennas" to attract nutrient molecules from the air? Some research indicates that they do. It has been found that the hairs on tobacco leaves collect large quantities of radioactive lead and polonium particles from the air, increasing the carcinogenic potential of tobacco smoke (*Nature*, 1974, vol. 249, p. 215–217).

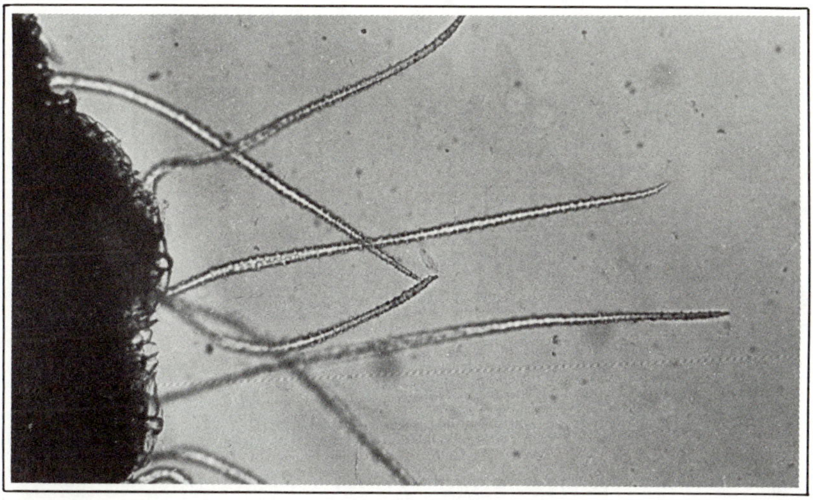

Alfalfa hairs (magnified 100 times under the microscope).

The ability of plants to absorb nutrients through their leaves is taken advantage of in the agricultural practice of foliar feeding, which can be a very useful and economical way to provide certain nutrients to crops, such as to supply trace elements which may be deficient or to obtain a quicker response from the plant than can occur by root absorption.

Plumbing. But we have wandered away from our subject. We were being carried along with the water and nutrient molecules into the tubular "plumbing" (xylem tissue) in the root of our plant. The *xylem* tissue of plants is one half of the plumbing system (or vascular tissue), the part which carries water and nutrients mainly upward in the plant. The other type of vascular tissue is *phloem*, which carries food and minerals from one place to another.

When water and nutrients get into the xylem, they can be quickly carried upward through the hollow, tubular vessels that form the "plumbing" by a sort of pulling force from above. Parts of the plant where water is needed (leaves and actively growing areas) create a water deficit, causing a pulling force on the water already inside the xylem vessels. So u-u-u-u-p we go!

Up the stem we rise. At every place on the stem where one or more leaves attach (called a node), there is a complex "traffic interchange," where the plumbing tubes of xylem and phloem divide into branches that go out into the leaves. As the growing season progresses, if a plant is not in good health, its "plumbing" can get plugged up in the nodes by mineral deposits, just as your house's plumbing can get clogged. As we will see later, this problem can be overcome by proper fertilization.

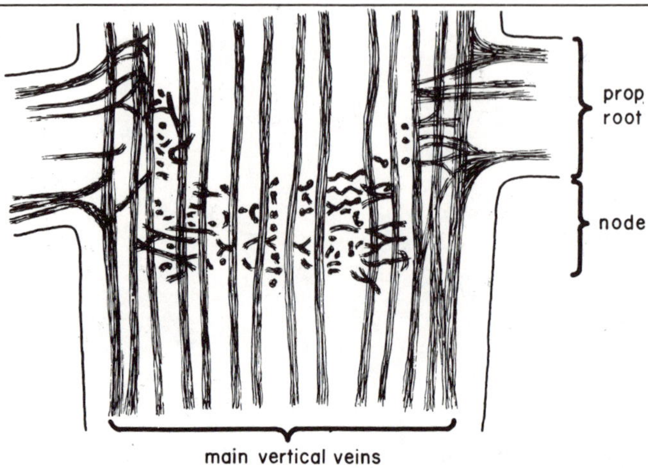

Thin vertical slice through a lower stem node of corn. Horizontal veins interconnect between vertical veins and between prop roots and other veins. Spots and dashes are veins that are turning corners and heading toward or away from the viewer.

The leaf. Up, up, up we go until we go through one of the nodal interchanges and zip out into a leaf. As we work our way along, the veins of the plumbing get smaller and smaller until we flow out the very end of a xylem vessel and find ourselves in a new and wonderful world. It is a beautiful place, something like being in the downtown streets of a large city, because we are surrounded by row upon row of tall, crystal-like columns, mostly transparent, but studded with many emerald green oval objects, the *chloroplasts*.

We have finally arrived, like Dorothy arriving at the Wizard of Oz's palace, at the FOOD FACTORY, the awe-inspiring place where the food is made which ultimately feeds nearly all life on earth (except for a few types of bacteria which can live on inorganic mineral materials).

Within the chloroplast-containing cells of any green plant, from lowly algae to majestic redwoods, a miracle occurs—called *photosynthesis*. This is the dry, technical name for the plant's chemically combining two common materials, water and carbon dioxide, into a life-sustaining molecule, sugar. In other words, the plant is a " chemical factory."

To understand more about this important process, we must digress to learn some more chemistry. There are different kinds of energy: light energy, heat energy, electrical energy, radioactive energy, and chemical energy. Our radios run on electrical energy, but our bodies "run" on chemical energy—all living organisms and cells do. Chemical energy is

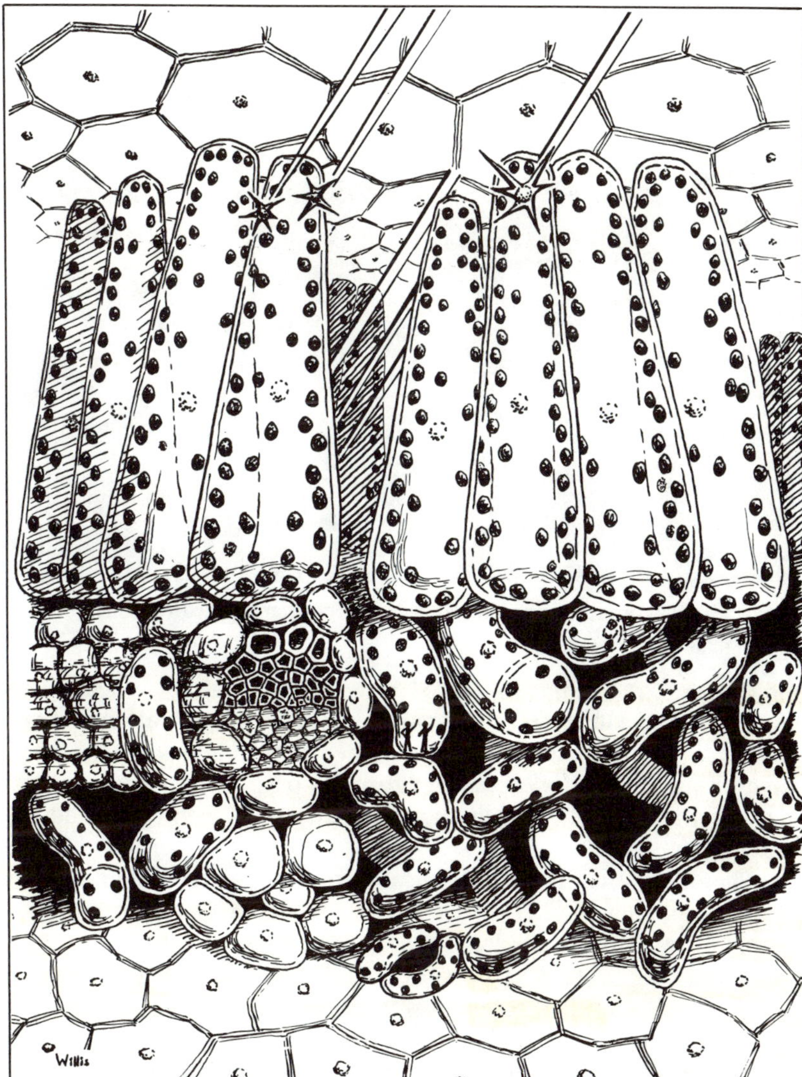

This view, drawn as if we were inside a leaf, shows the different parts. The upper epidermis, like a transparent skylight, lets light in. Beneath are the tall palisade cells where most food is made. The lower half of the leaf is mostly filled with loosely packed (to let air circulate) spongy parenchyma cells, where additional food is made. A vein (left center) supplies water and minerals and carries away food. At bottom center, in the lower epidermis, is an opening, the stoma, through which gases pass. Its size is regulated by two sausage-shaped guard cells.

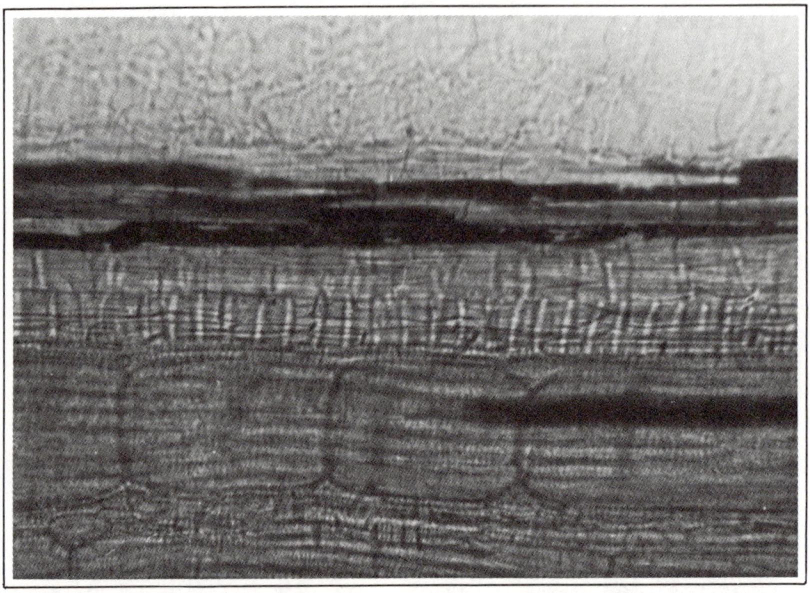

Xylem cells in a corn stem plugged with a dark deposit, magnified 100 times under the microscope.

the energy found in the various *chemical bonds* that hold together the atoms that make up a molecule or ion. When two or more atoms are held together, chemical energy is *stored* in their bonds, and when the bonds are broken, energy is *released*. To illustrate how much energy is stored in plant matter, just burn it; the chemical energy is then converted into heat energy.

Photosynthesis. In the "chemical factory" of a green plant cell, the raw materials of water (H_2O) and carbon dioxide (CO_2) are combined—this requires energy—to form the "manufactured product," a larger sugar molecule ($C_6H_{12}O_6$). The energy used to combine the water and carbon dioxide is light energy from the sun converted into chemical energy. Just as a factory assembles its product on an assembly line, the sugar molecule is not just put together in one step, but on a chemical "assembly line" in and near the green chloroplasts inside the cells. The chloroplasts are green because they contain the pigment *chlorophyll.* Chlorophyll is what "traps" (absorbs) the light energy from the sun. The sugar molecule (food) is thus a form of stored chemical energy. Incidentally, we run our automobiles on stored photosynthetic energy trapped thousands of years ago and changed into oil.

Some of the details of the assembly line of photosynthesis are shown in the accompanying diagram:

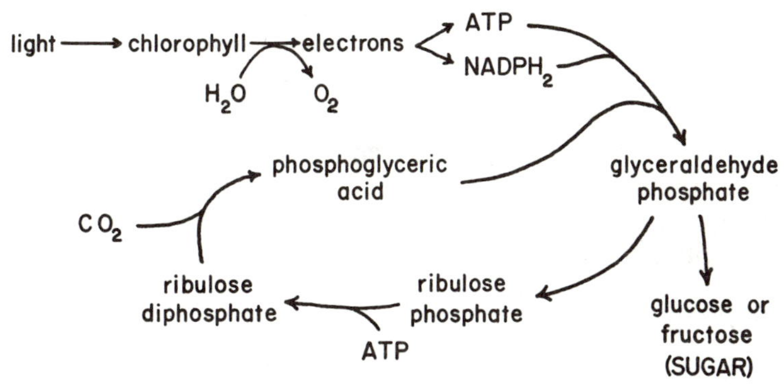

Four main chemical molecules are changed from one into the next in a cycle, with a molecule of sugar being "cranked out" at every revolution.

Two other components of the assembly line are vitally important: energy carriers and enzymes, for without these, the factory would grind to a stop.

Energy carriers. Chemical energy in cells is temporarily stored and carried from place to place in the cell by special energy carrier molecules. Some of the most common are ATP (adenosine triphosphate), $NADPH_2$ (nicotinamide adenine dinucleotide phosphate), and $FADH_2$ (flavine-adenine dinucleotide). These energy carriers contain one or more atoms of *phosphorus* as part of their molecule. AHA! Does that sound familiar—phosphorus? That's one of the macronutrients in plant nutrition! This is one of the main reasons plants (or animals) need phosphorus, to carry energy inside their cells. Phosphorus has been called the "energy currency" of plants (*Phosphorus for Agriculture*, Potash/ Phosphate Institute, 1978, p. 6).

Enzymes. Nearly all "assembly line" chemical changes in cells require enzymes, which are biological catalysts ("spark plugs"). Catalysts speed up the rate of chemical reactions. Otherwise the chemical reactions would not occur at a rate fast enough to sustain life. Enzymes are composed of a large protein molecule, plus usually another necessary part called a cofactor, coenzyme, prosthetic group, or metal activator. The cofactor can be either a metal ion (iron, copper, zinc, magnesium, potassium, or calcium) or an organic molecule (a vitamin or modified vitamin). Now do you see why some of the other plant nutrients, including some of the trace elements, and vitamins are so important in plant (and animal) nutrition?

There are thousands of enzymes in a single cell. Each kind of enzyme is very specific and can only affect one specific chemical change, one step of the assembly line. It is believed that an enzyme fits like a key in a lock on the surface of the substance or substances it controls (called the substrate). Actually, only certain parts of an enzyme, called active sites, are necessary for the enzyme to operate. The enzyme simply holds the reacting chemicals together long enough for the reaction to occur. An enzyme is not "used up" in the reaction; it is released to perform the process over and over. Enzymes can do a fantastic amount of work, from 10 to 10 million reactions *per second*!

Fertilize with CO_2?

Any required material which is in short supply or an improper condition that prevents a metabolic process from occurring as rapidly as it could is called a *limiting factor*. In the case of photosynthesis, inadequate light in the morning or evening or on a cloudy day would be a limiting factor, as would cold temperatures and lack of sufficient carbon dioxide (CO_2). The air only contains an average of 0.03% CO_2.

From an article by C. H. Wadleigh in the 1957 U.S.D.A. Yearbook of Agriculture, *Soils* (p. 41): "Under high light intensitites during the summer, this low level of CO_2 in the air [0.03%] probably is the main limiting factor in the photosynthesis. The production of an acre of a 100-bushel corn crop requires about 4 tons of CO_2. On a still summer day, the CO_2 level in the air of a cornfield drops very low. The corn under full sunlight could use 20 times the normal level of CO_2. We have evidence that the CO_2 produced by the respiration of micro-organisms in the soil is an important factor in the supply of the gas to photosynthesizing plants: A well-fertilized soil rich in decomposing organic matter provides a much higher level of CO_2 in the air just above the soil than does a barren, infertile soil—proof that soil management can affect photosynthesis in the plant."

From the Oct. 1968 *World Farming* (p. 31): "Agronomists and even some farmers in the U.S. are making high maize yields even higher by adding carbon dioxide (CO_2) to their bag of practices. Carbon dioxide is a basic requirement for plant growth. Plants combine it with water to form sugars. A farmer in Indiana, U.S.A., tried reinforcing the air surrounding his maize field with carbon dioxide on the premise that on sunny days plants well supplied with water and fertilizer cannot always find as much CO_2 in the atmosphere as they could use. Adding CO_2 to greenhouse atmospheres to increase yields is already an accepted practice. But fertilizing with CO_2 in field crops is definitely exploring new territory. The Indiana farmer used dry ice as his source of CO_2, dropping one-pound cakes every 25 feet during the middle of the growing season. The small section of the field treated with dry ice yielded a startling 211 bushels per acre, compared to 158 bushels per acre on untreated maize. This doesn't prove CO_2 fertilization is practical, but the results are encouraging further experimentation."

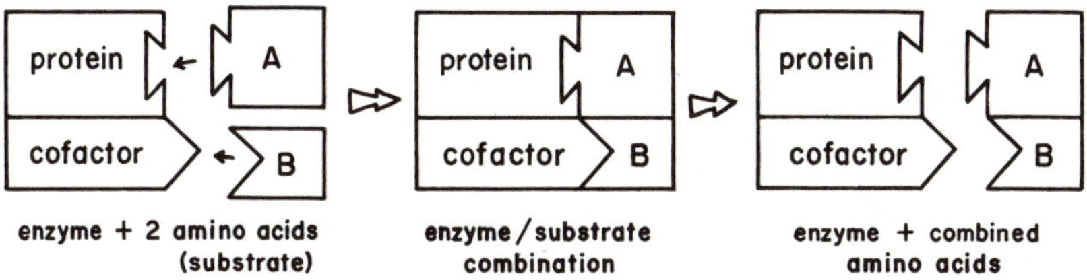

Respiration. The food produced by photosynthesis is used partly as a "fuel" to "run" cell activities, including some of the steps of photosynthesis. The stored chemical energy in food molecules is released in the energy-producing reactions of *cellular respiration*. During cellular respiration, the complex food molecules (carbohydrates, fats, and proteins) are gradually broken down in assembly line fashion (a series of chemical changes) to release the energy stored in their chemical bonds. The energy is then trapped and carried around by the energy carrying molecules mentioned before. The details of cellular respiration need not concern us here, but they are very complex, consisting of dozens of steps, nearly all of which require their own specific enzymes. How marvelously complicated is even one tiny cell!

Growth. The food produced by photosynthesis (glucose or fructose, called simple sugars) is used partly as "building blocks" to make new cells or cell parts, or to make special substances needed by the plant (such as enzymes, hormones, and pigments). Again, by a chemical factory assembly line series of changes, the simple sugars can be converted into more complex molecules, such as:

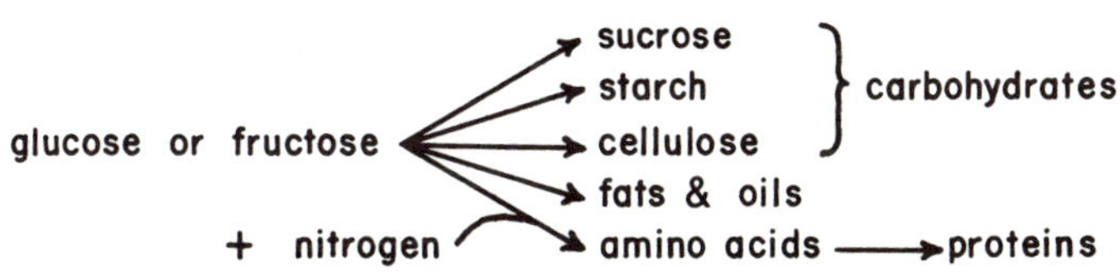

Again, specific enzymes are needed for the steps. Again, see the importance of trace elements and other nutrients in the soil?

Translocation. The transfer of food and the other manufactured products just mentioned from one part of the plant to another is called *translocation*. It is very important to the living, growing plant because some parts of the plant produce or supply materials needed in another part. For example, food produced in the leaves goes to all other parts, especially developing seeds, fruits, or tubers; and hormones produced in one area affect processes in another.

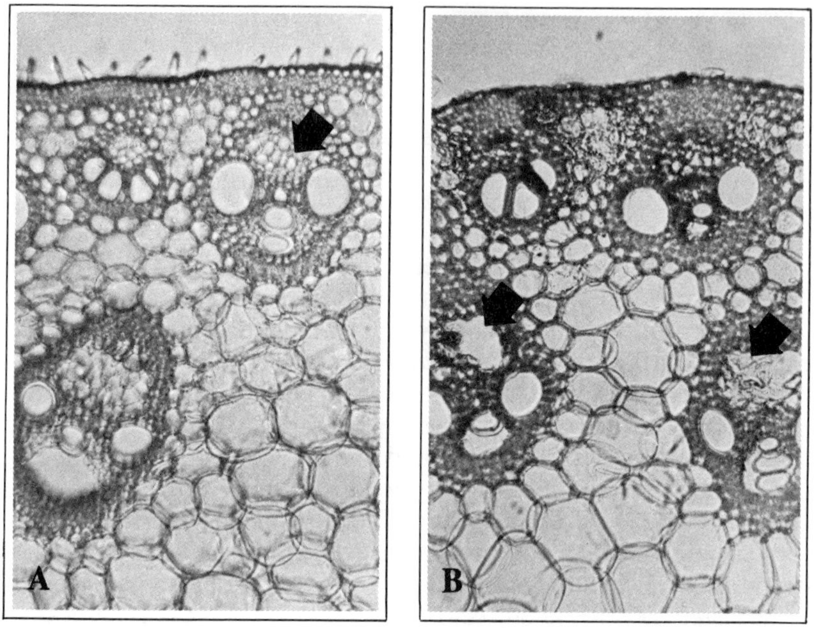

A. normal corn stem in cross section with several phloem cells (arrow) in each vein; B. corn stem with phloem disintegrating or missing (arrows) (magnified 100 times under the microscope).

Translocation over short distances can occur from one cell to its neighbor, but materials are mainly carried in that other half of the "plumbing," the *phloem* tissue. Just as the vessels of the xylem can become plugged up in an unhealthy plant, so also can phloem tissue, interfering with normal growth and seed or fruit production. Lack of available potassium causes corn nodes to become plugged by iron deposits; roots cannot receive adequate food and starve; yields are cut (E.J. Hewitt & T.A. Smith, *Plant Mineral Nutrition,* 1975, p. 171). So proper fertilization to maintain plant health is very important to producing high yields and high quality crops.

Hormones. One final digression before we complete our trip through the living plant. Most of us have a hazy realization that there are substances called *hormones* that regulate certain processes in animals and humans (such as growth and sexual functions), but few people know that plant growth and reproduction is also regulated by hormones, sometimes called growth substances. Hormones are chemical substances produced in small amounts in one part of the plant which move to other regions and act as chemical signals or messengers to regulate various processes. Known plant hormones include auxins, gibberellins, cytokinins, ethylene, and abscisic acid. They regulate such functions as cell division, stem-tip and root-tip growth and elongation, expansion of young leaves, fruit ripening, dropping of old leaves, and seed germination.

Sometimes man can use plant hormones or similar synthetic chemicals for certain purposes; for example, gibberellins may prove useful as growth stimulators in agriculture, while certain herbicides (2, 4-D and 2, 4, 5-T) are synthetic auxins.

Reproduction. But back to our trip. We started out being carried from the roots up to a leaf, where we witnessed the marvel of photosynthesis and other chemical transformations in the "food factory." Finally, let's follow a food molecule into a developing seed or fruit, the reproductive parts of a plant.

In order for seeds or fruit to be produced, flowers must be produced, and in order for flowers to be produced, flowers buds must first be formed. Tiny flower buds are formed long before they become obviously visible. In corn, the cob and tassel buds form when the plant is only about knee high. In apple trees, the buds that will produce next year's apples are formed this year. The number and quality of fruits or seeds depends partly on the number and health of the flower buds (plus other factors, such as weather, light, nutrients, and pollination). Some of these factors cannot be controlled by the farmer, but others can, especially the nutrition and health of the plant. Again, can you see why fertile soil is so vital?

After flowers have been pollinated, seed and fruit development begin. During this time, the nutrition and health of the plant are of utmost importance. There is a redistribution of water, sugars, amino acids, organic acids, inorganic nutrients, and hormones from roots, stems, and leaves to the developing seeds and fruits. The needs for some soil nutrients is at a peak at this time. Therefore the "plumbing" (both xylem and phloem) must be in good condition, as well as all metabolic and photosynthetic activities. In other words, the plant must be in excellent health. But as we have seen, both xylem and phloem can be plugged if the plant is not healthy. And some causes of poor health can be traced to the soil: insufficient nutrients, nutrients not available or not in correct form, improper balance of nutrients (various nutrients interact; too much of one compared to another can have adverse effects), and these situations can often be traced to insufficient humus and soil organisms.

So, from our place in the leaf, let's jump into a phloem cell and go gliding down the leaf, back into the stem, and out into a developing corn cob, for example. Here we are, and what a beehive of activity!

In earlier seed and fruit development, there is much cell division,

producing new cells; later the existing cells simply enlarge as the seed or fruit swells. A great deal of food and other chemical substances from the leaves are moving into the seed or fruit for storage. Later, in the grain crops, much of the moisture in the kernels is (or should be) removed, producing a "dry" grain which can be stored without getting moldy. In both seeds and fruits, there are also chemical transformations in the type of stored food, from sugars to starches, organic acids, proteins, and fats or oils.

All of these metabolic activities require cell energy (supplied by phosphate-containing energy carriers), enzymes, and often hormones. And to occur normally and produce top-quality seeds or fruits, the plant must be healthy, with adequate water and nutrients. If there are trace element deficiencies for example, certain enzymes which regulate vital metabolic steps may be deficient, leading to low test-weight or low bushels per acre or low quality (low biological value) of animal feed. Or if the all important "plumbing" is plugged, needed food may not enter or water may not be removed to "dry down" grain naturally, requiring costly grain drying. These problems can largely be prevented if the soil is "healthy," with good tilth, adequate humus and soil organisms, and proper fertilization.

So here we are in a plant's seed or fruit. CHOMP! Oh, oh–look out, we're being eaten by an animal!

CHAPTER 5

Top of the Pyramid— Animals

The food chain is often diagrammed as a pyramid, called the biotic pyramid, with plants on the lower level, plant-eating animals (herbivores) next, and animal-eating animals (carnivores) next. If man is included, he is placed at the very top. The reason a pyramid design is used is to show that the numbers of species and individuals are greater at the lower levels and smaller at the higher levels of the food chain, because from one level to the next (when one organism eats or is fed by another), there is a certain amount of energy loss or wastage (food is stored energy).

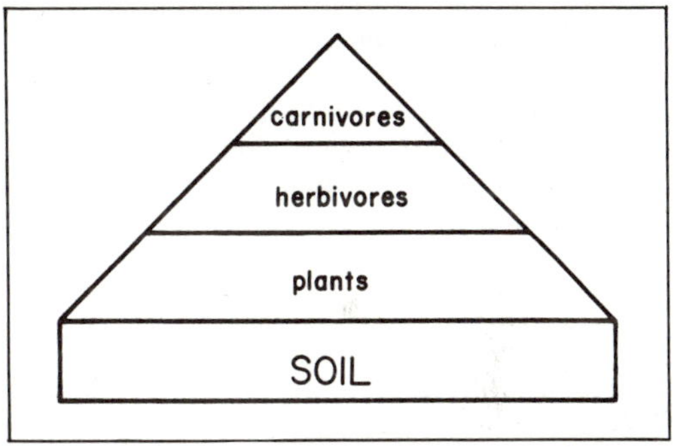

So when we move from the plant realm to the animal realm, we are moving to the top of the biotic pyramid. What affects animal (and human) health and nutrition? Are animals really that much different from plants? If you raise livestock or if you simply want yourself and your family to be healthy, you should know something about animal (and human) nutrition.

Sickness. Is it normal to be sick? Some people are ill so much of the time that they have forgotten what it was like to be well. Most farmers expect to have a certain number of sick animals and young ones that die. It doesn't pay to be sick. Doctors or veterinarians, drugs and medicines don't come cheap. In 1980, Americans spent $247 billion for human "health care" (or is it really sickness care?), which is 9% of the gross national product. Think how much money and time could be saved if you or your animals were never sick.

Unfortunately, medical and veterinary colleges spend little time teaching future doctors how to keep their patients healthy, how to *prevent* disease. They almost entirely follow the logic of diagnosing a disease, giving it an unintelligible name, and prescribing some drug or poison to fight it, often by attacking the symptoms, not the real cause.

Sickness is NOT normal! Robust *health* is the normal condition of animal and human bodies. The dictionary defines disease as "disturbed or abnormal structure or physiological action in the living organism," and sickness as "any disordered and weakened state."

We often hear the phrase, "you are what you eat," and the Bible reaffirms it, "All flesh is grass." By now you know that animals get their food, which later becomes their body substances, from either eating plants or from eating animals that eat plants. And where do plants grow? In the soil. Dr. William A. Albrecht said it very well in his book, *Soil Fertility and Animal Health* (1958):

> "We know that the soil grows grass; that the grass feeds our livestock; and that these animals, in turn as meats, are our choice protein foods. We can therefore, connect our soil with our health via nutrition. Since only the soil fertility, or that part of the soil made up of the elements essential for life, enters into the nutrition by which we are fed, we may well speak of animal health as premised on the soil fertility.
>
> "There are increasing reports that animal afflictions are coming to be viewed as sins of omission, more than that our livestock is falling prey to some stealthy force (possibly microbes). It is obvious that if an animal is deprived of some essential body constituent, troubles in its health will be encountered."

Animal nutrition. In some ways, animal nutrition is the opposite of the nutrition of plants which we saw in Chapter 4. Remember that plants take simple chemical substances (carbon dioxide, water, and minerals) and combine them into more complex substances (carbohydrates, fats, proteins), trapping and storing the sun's energy in the form of chemical energy. Animals, in their body and cell functions, first of all do the reverse: break down the complex food molecules, releasing some of the stored energy; but then they do as plants do, namely combine the simpler "building blocks" obtained from food substances into more complex substances that make up the cells and tissues of the animal's body. And you may recall that plants also can break down complex food molecules into building blocks and energy. Amazingly enough, many of the basic molecules involved in animals are nearly or exactly the same as those found in plants; such as amino acids, ATP, vitamins, etc., but many others are different, since plants do not have muscles, nerves, blood, etc.

So let's take a quick tour through an animal's body and follow a bite of food from a plant to see what happens to it. First an overview of what we will see: food is (1) digested in the animal's digestive tract, (2) absorbed into the circulatory system, (3) transported to all cells, (4) completely broken down to obtain energy or else assembled into new cell components, and (5) wastes are eliminated. Another complication in agriculture is that some animals have digestive systems pretty much like man's (pigs, horses, poultry), but others, called ruminants, have an unusual system with special features of the stomach (ruminants include cattle, sheep, and goats). We will first discuss non-ruminant (and human) nutrition.

Digestion. In most animals, food is first chewed in the mouth, which breaks it up into smaller particles to facilitate digestion and mixes it with saliva. Saliva not only makes the food easier to swallow, but in some animals also contains an enzyme that digests starch, called ptyalin. Upon being swallowed, we descend down the narrow esophagus and enter the large, sac-like stomach. Contrary to popular belief, less digestion takes place in the stomach than in the small intestine. In the stomach, the semi-liquid food-saliva mixture is churned and mixed with gastric juice, which contains strong hydrochloric acid and a few enzymes that act on proteins and fats. Preliminary digestion occurs, and the food is then pushed out into the long tubular small intestine, where most digestion occurs, plus absorption of digested food into the circulatory system. About 10 enzymes produced in the small intestine finish breaking down the food's carbohydrates, fats, and proteins. A summary of digestion is as follows:

starches, complex sugars → simple sugars

fats → fatty acids, glycerol, mono- & diglycerides

proteins → amino acids

Undigestible materials move into the large intestine, where some water is absorbed to conserve body water. The large intestine also contains a multitude of bacteria, some of which are very beneficial in that they produce vitamins used by the animal's body, and in some cases they digest cellulose in similar fashion to ruminants' bacteria. Finally the waste matter, called feces, is eliminated.

Ruminants. Now let's consider those marvellous animals, the ruminants: cattle, sheep, and goats (plus members of the deer and camel families). These are the animals that "chew their cud." Their "stomach" is equipped with four compartments: the rumen ("paunch"), reticulum ("honeycomb"), omasum ("manyplies"), and abomasum. The first three are really expansions of the lower esophagus (called the forestomach), while the abomasum represents the true stomach, as compared to non-ruminant animals. A ruminant eats a large amount of plant food rapidly, swallowing it into the rumen. Then it is regurgitated and rechewed ("chewing the cud"), and reswallowed into the other stomach compartments. In the rumen and reticulum are billions and billions of special bacteria and yeasts, plus lesser numbers of protozoans. These microorganisms, especially the bacteria, are vital for the normal nutrition of a ruminant animal. First they digest the cellulose part of the plant-derived food (from the plant's cell walls), which most animals cannot digest, breaking it down and releasing by-products of carbon dioxide, methane, and fatty acids, the latter being used by the ruminant as a source of nutrients. Also, the bacteria produce essential amino acids and many vitamins (those classified as water soluble vitamins) for the animal's use. They can also turn an inferior type of protein (lacking some amino acids) into a higher quality food source by changing one amino acid into another. These nutrients become available to the animal when the microorganisms are digested by the later parts of the digestive system. Thus, because of their microscopic helpers, the ruminants are virtually self-sufficient and can live on very poor quality feed. The use of antibiotics can kill or upset the balance of the microbes in the rumen; this can lead to stress, sickness, and even death.

Absorption. But back to the digested food molecules (simple sugars, amino acids, fatty acids, glycerol, and glycerides), plus the other needed food materials of water, vitamins, and minerals. In order to be used by the body, these substances must enter every living cell, which means they must be absorbed from the digestive system (stomach, small and large intestine) into the blood and lymph of the circulatory system, and transported throughout the body, where they can enter whichever cells need them at the time. Any excess or unneeded food molecules can be either stored for future use or excreted as waste. The rate of absorption of one element is sometimes influenced by another; for example, a large amount of phosphate interferes with calcium absorption.

The efficiency of absorption by the small intestine is greatly increased by its millions of tiny fingerlike folds (villi) and their billions of microscopic microvilli, which add surface area to the intestine. It has been discovered that feeding animals harsh chemicals or drugs and antibiotics can destroy the microvilli and thus render the animal a "digestive cripple" for the rest of its life. This simply adds another stress to the body, which leads to general poor health or disease.

The products of carbohydrate and protein digestion (simple sugars and amino acids), and small fatty acids, are absorbed through the stomach or mostly the intestinal wall into the blood stream, but most of the products of fat digestion are first absorbed into the lymph and then flow into the blood.

Cellular metabolism. The blood carries needed materials—digested food, vitamins, minerals, water, oxygen, hormones—to or near to every living cell. They enter the cells and are used in the many marvellous activities of living cells. These can be boiled down into two categories: (1) catabolism, or cellular respiration, in which the digested food molecules are further broken down to release the chemical energy stored in them, and (2) anabolism, or assimilation, in which digested food molecules are used as "building blocks" and reassembled into new cells or cell parts, or into useful secretions which some cells produce, such as enzymes, hormones, mucus, and milk.

All of the cell activities, both catabolism and anabolism, are often called metabolism, or metabolic activities. As we have seen for plants in Chapter 4, these metabolic activities, both breaking down and building up, proceed in assembly line fashion, in many steps, most of which require their own specific enzymes in order to occur at a sufficiently rapid rate. Again, the activities of metabolism are very complex, and we will not go into any detail here.

It is interesting to note that most of the steps of catabolism are identical in plant and animal cells. The energy released from food molecules can be used to "run" the various activities that any cell performs, including those of anabolism, or it may be used for the special job of certain cells, such as muscle and nerve activity. As in plants, the released energy is temporarily trapped and stored in energy carrying molecules, primarily ATP (remember, they contain phosphorus).

Hormones. In animals, the metabolic functions of cells are regulated by certain hormone-secreting glands: (1) the thyroid and parathyroid glands, which regulate general cell metabolism, plus the body's use of calcium and phosphorus; (2) the pancreas, which regulates the body's use and storage of sugar and starch (a malfunction of this gland can produce the well-known malady of diabetes); (3) the adrenal glands, which control the metabolism of minerals, water, carbohydrates, fats, and proteins, as well as preparing the body to exert extra energy in situations of fear, danger, etc.; and (4) the pituitary and hypothalamus of the brain, which control the other hormone-producing glands.

Excretion. The final function involved in nutrition is the excretion, or elimination, of waste products. These include undigestible materials in the food eaten, digested materials that the body cannot use, and the by-products of metabolism, which are mainly nitrogen-containing urea (uric acid in birds) and carbon dioxide. These wastes are eliminated through four pathways: the intestine, the kidneys, the lungs, and in some animals, the skin.

Health. So we have completed our trip through an animal's body, following what happens to food molecules, or what should happen, for what we described is the *normal* functioning of the body and its cells. But just as plants can be unhealthy and have disturbed metabolism, partly from improper nutrients from the soil, so also can animals become sick, partly from improper nutrients in the food they eat. We humans certainly should realize that eating the wrong things or not getting enough vitamins and minerals will lead to health problems—the same thing is true of our livestock.

To be sure, other factors than feed can cause health problems in farm animals. One could say that sickness is caused by the body's succumbing to one or more stresses. Stresses can come from many sources: poor ventilation, impure or contaminated water, adverse weather conditions (temperature, humidity, precipitation, wind), infectious diseases, parasites, improper handling and injury, noise or other frightening conditions, lack of exercise, and of course improper nutrition. Most of these stresses can be reduced or eliminated by good farming practices and good animal management methods. Anything that we can do to reduce stresses will pay off in increased meat, milk, egg, or wool production, and reproduction of healthy young.

We cannot spend time on most aspects of animal management. What we are most interested in is good nutrition and how it relates to plant growth and the soil. For, just as is true for plants, an animal that is well nourished is better able to resist or tolerate such stresses as diseases, parasites, and adverse weather.

Nutrients. So let's back up and look in more detail at the nutrients needed by livestock and poultry. Remember, what are the two reasons animals need food—or to what two uses is digested food put in the cells? First, energy production, and second, building new cells, or growth and reproduction, plus repair of injury, and milk or egg production. Now let's look at the basic nutrients and see what their functions are. For energy production, the animal normally uses carbohydrates and fats as "fuels," although in starvation or excess, proteins can also be used. For growth, the necessities are proteins and minerals (calcium and phosphorus for bones and teeth, iron in the blood, sulfur in certain amino acids). But the cellular metabolic activities of both energy production (catabolism) and growth (anabolism) also require thousands of different enzymes, and as we covered in Chapter 4, enzymes are made of a protein plus a cofactor, which is either a mineral element or a vitamin. So here we see the need for vitamins and certain minerals, such as copper, zinc, iron, calcium, potassium, and magnesium. And then, certain minerals have other vital functions: phosphorus is part of the energy carrier molecules, sodium and potassium regulate internal osmotic pressure (water balance), chlorine and potassium regulate pH, iodine is part of the thyroid hormone, and so on. Finally, water is an essential nutrient substance since it is a necessary component of all living cells; in fact cells are mostly water. An average animal cell is about 70% water.

Requirements. Some nutrient substances are essential in an animal's diet, while others can be manufactured within the animal's body or cells. The following mineral elements are essential: calcium, phosphorus, sodium, potassium, chlorine, magnesium, iron, manganese, copper, zinc, fluorine, sulfur, iodine, and cobalt. About 10 or more of the amino acids are essential for most animals. Vitamins are essential. However, in ruminants and most plant-feeding animals, some amino acids and vitamins are provided by the bacteria living in the digestive tract. Most animals, other than man, monkeys, and guinea pigs, can manufacture their own vitamin C.

But it isn't as simple as being sure that an animal's diet contains the essential nutrients, for the nutrients must be present in the proper amounts and proportions to produce optimum good health. For example, a 1500 pound lactating cow may need 21 grams of calcium per day, but only 0.1 milligram of cobalt. Some elements are toxic in too large amounts, such as fluorine. There are interactions between some elements; for example in dairy cows, a calcium-phosphorus ratio greater than 2:1 can lead to milk fever.

Quality. So we see that for good animal health, not just any old food will do—we need *high quality food*. Food that will provide the right amount of the right kinds of nutrients in the right proportions. Are most livestock being fed really high quality food directly from the fields today? No, today we see the burgeoning practice of supplying feed supplements and trace mineral additives, because most farmers are not growing high quality crops. Their crops are incomplete, out of balance, and in some cases, even toxic.

How can we measure and evaluate the nutritional value of livestock feed? There are several methods.

Nutritional value. Three basic methods are used to determine *total energy* value of food. One is the total digestible nutrients, TDN. This is found by feeding a certain feed to an animal, collecting its feces, and finding the difference between the amount of nutrient fed and the amount in the feces. A "*digestion coefficient*" is calculated for major nutrients, including protein, nitrogen-free extract (sugars, starches, nonnitrogenous organic acids), fiber, and fat, as follows:

$$\text{digestion coefficient} = \frac{\text{intake} - \text{fecal excretion}}{\text{intake}} \times 100 = \%\,\text{digestibility}$$

Then the percent of each nutrient in the feed is multiplied by its digestion coefficient and the figures for each nutrient are totaled to give an overall % TDN. The use of TDN to measure food quality has been around for a long time, but has some disadvantages. It does not work well for ruminants, and it does not measure all energy losses from the body, only the feces.

A better method of measuring food value is the caloric system. A calorie is a measure of heat energy; technically, it is the amount of heat required to raise the temperature of one gram of water one degree Celsius (1.8°F). The heat energy value of any food can be measured by placing a sample in a container called a calorimeter and actually burning it. The temperature rise of a water bath around the calorimeter can be easily measured and the calories calculated. Generally the number of calories in food is large, so the "large calorie," or kilocalorie (1000 calories) is used. It is sometimes written Calorie, with a capital C. The total energy given off when a food is burned in the calorimeter is called the gross energy. But this does not measure the amount of food energy the animal can actually use from the food. Again, a digestion trial is run and the animal's feces collected and analyzed. From this, *digestible energy* can be calculated: DE = gross energy − fecal energy. Digestible energy represents the part of the food energy which is absorbed from the digestive tract. But this is not the energy able to be used by the animal's cells, because some energy is also lost through the urine and from gases generated by rumen bacteria (mostly methane). So we can analyize the animal's urine and exhaled breath by confining it in a special cage, and then calculate *metabolizable energy:*

$$\text{ME} = \text{DE} - (\text{urine loss} + \text{gas loss}).$$

The Calorie food values used in human nutrition are similar to metabolizable energy.

But metabolizable energy does not consider the amount of body heat lost by the animal. This can be measured by confining the animal in a large calorimeter. Finally, we can calculate the *net energy*:

$$\text{NE} = \text{ME} - \text{heat loss}.$$

This method is the most precise way of determining food value. A recent article in *Hoard's Dairyman* ("Some Common Mistakes in Feeding Dairy Cows," Feb. 10, 1982) states that a high quality forage should contain 60 therms of net energy milk per 100 pounds of dry feed (1,000,000 calories = 1 therm). "Trying to get milk out of anything but high quality forages is futile and expensive."

A third method of determining food value is by a comparative slaughter-feeding trial. In this method, a number of animals are slaughtered before the trial to determine their body composition (either by chemical analysis or simply by finding the density of the carcass by weighing it in water). At the end of the feeding trial, the remaining animals are slaughtered and analyzed. The difference between the two groups of animals is called the *energy storage*. This method has the advantage of allowing the animals to be fed under more natural conditions than those in the calorie method.

Protein value. Another measure of food value is the *biological value*, BV. It is not a direct measure of the *total* food value, but a measure of protein quality. It compares the relative proportions of essential amino acids in the food with the animal's needs. A protein that lacks any essential amino acid is a poor quality protein. Since proteins contain nitrogen, BV is actually calculated as

$$\frac{\text{nitrogen retained}}{\text{nitrogen absorbed}}$$

In other words it measures the nitrogen retained for the animal's growth, reproduction, or maintenance, which is determined by urinary and fecal analysis. Ruminants are special in that they have few amino acid requirements, since rumen bacteria can make amino acids from non-protein sources of nitrogen. Only cysteine and methonine are essential amino acids for ruminants.

Other measures of protein value of food are used, especially *crude protein* (CP) in agriculture. The name is deceiving because actually protein is not measured, but the food is "burned" to ash with strong acid and the percent of total nitrogen found. This is then multiplied by 6.25 to give percent crude protein. This fairly simple method of finding food protein value is, as its name suggests, a *crude estimate*, since it does not consider the balance of amino acids nor the essential amino acids, and it is based on the *assumption* that the food nitrogen compounds are mostly the amino acids which make up proteins. But this may not always be the case, depending on the nutritional history of the food. In other words, there can be non-protein nitrogen (NPN) compounds in the food, such as amides, and a single nitrogen atom in the amino acid tryptophan, which non-ruminants cannot utilize to build proteins (except for horses, which have bacteria in their cecum [comparable to our appendix] similar to those of ruminants). Also, if plants are grown in soil having an excess of nitrates, as commonly happens when farmers apply too much inorganic salt fertilizers, the plant can contain a great deal of nitrates, which can even be toxic to animals.

Food quality. Thus we see that the *quality* of food, not just bulk, is extremely important in promoting animal health. And one of the most important aspects of food quality is the amounts of essential amino acids. Laboratory tests have shown that plants grown in mineral deficient soil have an unbalanced amino acid production. In a test using peppermint, plants grown in calcium-deficient soil produced too much asparagine and proline, and too little glutamine, glutamic acid, and aspartic acid. Plants grown in sulfur-deficient soil produced too much asparagine and arginine, and too little glutamine, alanine, serine, glutamic acid, and aspartic acid (see box, Chapter 17).

But there is more than that. Nutrition tests of rats and chickens have shown that animals fed mixtures of pure, chemically synthesized amino acids do not grow as rapidly as those fed high quality natural proteins. So there are other nutritional substances present in natural foods which are needed for optimum health.

Even though ruminants and horses do not need such high quality protein as other animals do, the young of ruminants do need high quality protein to supply the essential amino acids because their rumens have not yet become activated.

Food efficiency. The "bottom line" of feeding agricultural animals is their efficiency in converting food into meat, milk, eggs, or wool. This goes beyond digestibility and assimilation into new body cells. The farmer is paid according to pounds and bushels. Quantity should not be the only determining factor in prices; quality should be considered too, but except for limited cases, that is how the system works.

Different species of animals vary considerably in their food efficiency; for example a dairy cow, the most efficient farm animal, is 25% efficient in using food calories and 34% efficient in converting proteins into milk. A layer hen is 10% efficient for calories and 16% for proteins in egg production, while a beef steer is only 3% efficient for calories and 9% for proteins in producing meat. But there are other factors involved: the particular breed or variety of animal, presence or absence of the various stresses listed earlier (weather, disease, parasites, etc.), the genetic variation of different individuals ("ol' Bossie is just a better producer than Daisy"), and of course the *quality* of the food the animal eats.

A small increase in food quality, whether measured as digestibility or biological value, can cause a much greater increase in food efficiency and agricultural production, and in overall farm efficiency. For example, dairy cattle fed high quality forage gives more milk, with a higher butter-fat content, AND *eat less feed* while doing it! That's a situation most farmers would certainly like to be in, but sadly, most aren't. Why? It all goes back to the soil and proper fertility for the plants grown in the soil.

So, to complete our journey through the ecological food cycle, we make one more visit to the soil, to close the circle.

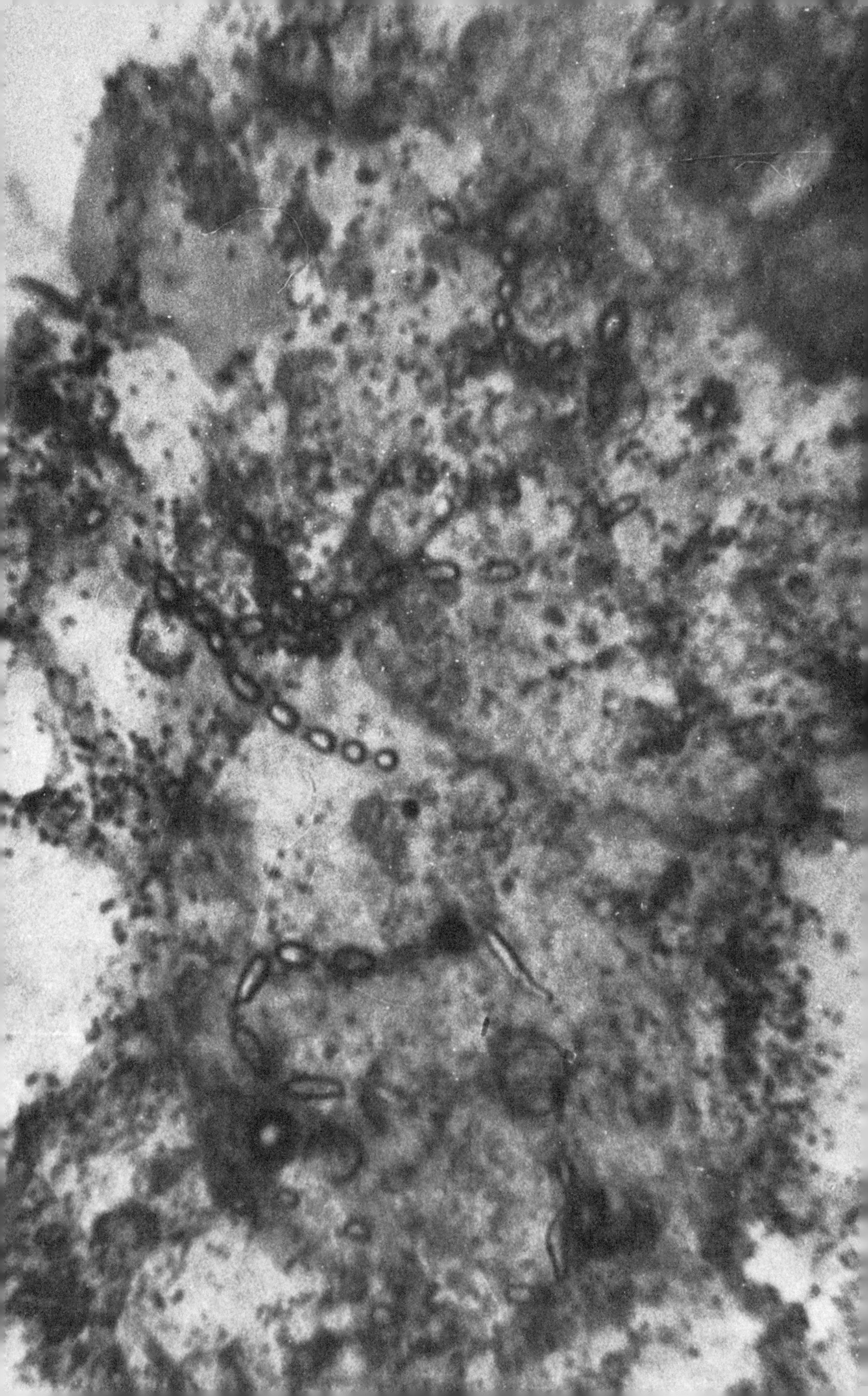

CHAPTER 6

Closing the Circle— Decomposers

Well, here we are down in the soil again. If it seems like we've been going in circles, you're right, we have! Remember, we are taking a trip through the food *cycle*. We now have to fill in the gaps and complete the circle.

In order to complete the food cycle, dead organic matter—dead bodies of plants and animals, plus the waste excretions of animals—have to be broken down and returned to the soil so that the nutrients they contain can be reused by plants—or to use the modern environmentalist's term, they have to be *recycled*.

There is a whole host of special organisms, the decomposers, who are ready, willing, and able to break down—to digest—organic matter. In doing so, they use it for food to build their own bodies or cells, but they later die or are eaten by other soil organisms, and eventually the nutrients are useable by plants.

We briefly met these denizens of the soil in Chapter 3. To review, they include both plants (bacteria, actinomycetes, fungi, and algae) and animals (protozoa, nematodes, earthworms, and arthropods, especially insects and mites). Some of them eat dead organic matter directly (bacteria, actinomycetes, fungi, earthworms, and some arthropods), while others are predators on them or each other (protozoa, nematodes, and most arthropods). Algae make their own food by photosynthesis like most plants. Some bacteria and fungi attack living organisms and may cause diseases and death of their hosts. A few bacteria can live entirely on inorganic nutrients, such as iron or sulfur compounds.

So in the soil we have a whole "subcommunity" or "mini-ecosystem" of producers, consumers, and decomposers. We are mainly interested in the decomposers, and the most important are the microorganisms: bacteria, actinomycetes, and fungi. In addition, the earthworms do perform valuable services if they are present. They consume fresh organic matter and help "chew" it into finer particles which can later be more readily attacked by the microorganisms. Earthworms "haul" much surface organic matter down to deeper layers of the soil. They enrich the soil by their wastes and dead bodies, and they aerate and turn over the soil by their burrowing activities.

But now let's concentrate on the smaller decomposers, bacteria, actinomycetes, and fungi. We briefly described their characteristics and activities in Chapter 3. Bacteria are single-celled and can live in either aerobic (with oxygen) or anaerobic (without oxygen) conditions, but generally cannot live in very acid conditions. They help decompose organic matter to form humus, convert inorganic chemicals into useful plant nutrients (including nitrogen-fixation), and detoxify (break down) man-made toxic chemicals.

Actinomycetes are filamentous (thread-like) and require aerobic, neutral or slightly alkaline conditions. They aid in producing humus.

Fungi are filamentous, mostly aerobic, and can tolerate more acid conditions. Some help produce humus, but one type, the mycorrhizae,

are symbiotic in plant roots, where they aid in nutrient absorption and root growth.

But now that that we've met the "work force," let's follow what happens when fresh organic matter decomposes. Most of the dead organic matter used in agriculture is of plant origin, either crop residues (dead stalks, leaves, roots, cobs, etc.) or animal manures (containing a large percent of partly digested plant matter). In recent years, some people have been using sewage sludge and papermill waste. So we are starting out with a food source for our hungry volunteer army containing a large amount of the components of plant cell walls: cellulose, hemicellulose, and lignin. These materials are very resistant to decay, particularly lignin, which can take months or years to break down. The rest of the plant material consists of whatever cell contents used to be in the plant, including sugars, starches, fats, and proteins, all of which are more easily digested. These initial components of organic matter have a high percent of carbon, compared to nitrogen, or a high carbon to nitrogen ratio,

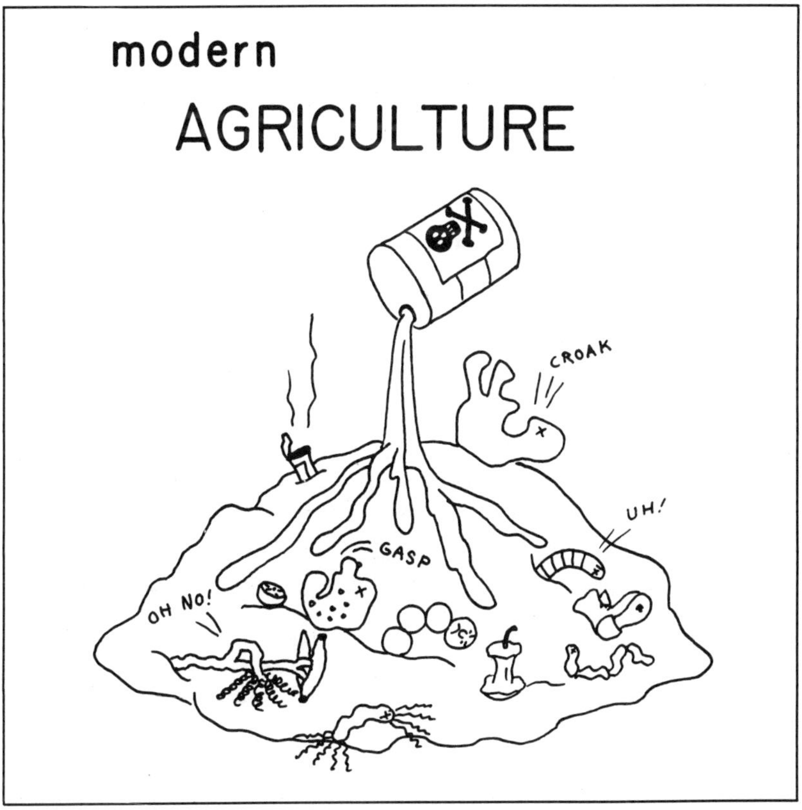

Earthworms

Earthworms, often called "nature's plowmen," render a valuable service to man in helping to enrich soil. "Earthworm" is one of many common names used for various land-living segmented worms (Annelida); different varieties are often called angle worms, fish worms, night crawlers, and red worms. They are not at all related to the nematodes, or roundworms or threadworms, which are smaller and also found in the soil. There are giant earthworms in Australia, Africa, and South America which can get up to 3 feet long (think of the fish you could catch with one of those!).

Earthworms are scavengers, feeding on dead or decaying organic matter, either by consuming it directly or by eating their way through the soil and extracting the small amount of organic matter in the soil. Their excretions are either plastered along their burrow walls or deposited on the surface as "castings." Analysis of earthworm castings reveals that they are richer in plant nutrients than the soil (about three times more calcium and several times more nitrogen, phosphorus, and potassium (K. P. Barley, *Adances in Agronomy*, vol. 13, 1961, p. 257). So along with the soil's army of microorganisms, earthworms help decompose organic matter and make the nutrients available to plants. Their "chewing" of fresh organic matter exposes more surface area for later microbial decomposition.

But that isn't all they do. Their burrowing activities increase soil porosity, aeration, and drainage, plus churning and mixing the soil. To see how they can mix soil, try placing several earthworms in a container that has the bottom half filled with sand and the top half with dark soil. In several months the sand and soil will be thoroughly mixed. Earthworms burrow several feet deep and can break up a hardpan layer.

Tests have shown that crops grown in earthworm-inhabited soil increased yields from 25% to over 3 times more than in earthworm-free soil (K. P. Barley, *Advances in Agronomy*, vol. 13, 1961, p. 262-264).

Earthworms are sensitive to unfavorable soil conditions. They cannot live in strongly acid soils or in too dry or waterlogged soils. They are killed by strong salt fertilizers and some pesticides (however, they can tolerate others, including DDT, but then birds are killed when they eat contaminated earthworms *(Pesticide Reviews*, vol. 57, 1975, p. 40-41). In general, earthworms can only do well in fertile soil, but proper use of inorganic fertilizers can build up soil humus (by producing abundant crop residues) and allow earthworms to become common. Dr. William Albrecht illustrated this by telling of a lawn around a new house that was seeded on "raw" clay with a thin layer of topsoil added. Ordinary N P K fertilizers and trace elements were added. Seven years later the lawn was dug up and there were numerous worm burrows, even down into the clay (The Albrecht Papers, 1975, p. 232-233).

An abundant earthworm population is not only a sign of fertile soil—they are busily at work to make it even more fertile. If you don't believe it, prowl around your garden or lawn at night with a flashlight after a rain!

from 20:1 to 80:1 or sometimes more. Animal manures not only contain considerable undigested food, with the above constituents, but also millions of bacteria, both living and dead, from the animal's digestive tract, plus other miscellaneous substances, including vitamins and important mineral elements: calcium, boron, manganese, copper, and zinc.

Oxidation-mineralization. What exactly happens as dead organic matter decomposes depends on circumstances. If the material is left exposed to the air on the surface, where temperatures are not favorable for microbial growth and moisture is too little, a slow process of decomposition occurs called oxidation-mineralization. The elements of carbon, oxygen, hydrogen, and nitrogen, which make up carbohydrates, fats, and proteins, are released into the air as gases: carbon dioxide, water vapor, nitrogen, and nitrogen oxides. The inorganic mineral elements—sulfur, potassium, phosphorus, etc.—are returned to the soil. This process of oxidation-mineralization is very wasteful, for much potentially useful energy and nutrients are lost into the air; only the minerals are saved for later plant use. However, some advantages of leaving organic matter on the surface as a mulch do exist: protection from erosion, better water absorption and retention, moderating soil temperature and weed control.

A much better use of nutrients can be made if the organic matter decomposes *in* the soil, not on it, for there the conditions of moisture and temperature are more favorable for the growth of microorganisms.

Microbial decomposition. There are two ways in which organic matter can decompose by the action of microorganisms: (1) anaerobic decomposition, or fermentation or putrifaction, and (2) aerobic decomposition. Anaerobic decomposition will occur wherever the organic matter does not contain much or any oxygen, such as inside a manure pile or deep in the soil (most soils are anaerobic below about 3-5 inches). As we shall see, anaerobic decomposition is also very wasteful of nutrients, or even toxic to plants.

Anaerobic decomposition. Typically, when organic matter decomposes anaerobically, it goes through two stages: (1) a brief period (about one week if moisture levels are adequate) called ammonification, during which certain bacteria release considerable nitrogen in the form of ammonia, NH_3. This causes the pH to rise into the very alkaline range (up to pH 8-9.5). (2) The second stage is called fermentation and lasts one or two months or more. More gases are released by bacteria and fungi: carbon dioxide, volatile organic acids, aldehydes, and alcohols come from carbohydrate breakdown; while methane, hydrogen sulfide, mercaptans, skatole, indole, amines, aldehydes, and others come from protein breakdown. Many of these have strong odors and cause the disagreeable odors associated with manure and putrid grain or silage. Some are toxic to plants and animals. The acids produced during this stage of decomposition lower the pH to 7 - 9. There is little cellulose or lignin breakdown in anaerobic decomposition (peat forms under anaerobic conditions). Anaerobic decomposition proceeds slowly and releases little heat compared to aerobic decomposition.

Aerobic decomposition. Under aerobic conditions, decomposition proceeds rapidly and releases considerable heat. Several types of microorganisms do the job, each taking a turn in breaking down the components of the organic matter. First, if temperatures are cool, certain fungi and bacteria called cryophilic ("cold-loving") bacteria (active at temperatures up to about 65°F), begin the work of breaking down carbohydrates (sugars, starches), fats, and proteins. The carbohydrates are mostly just "burned" for energy, with carbon dioxide and water being produced as by-products. Nitrogen from proteins and their amino acids is used to produce more bacterial cells. Heat is also produced as a result of metabolism.

Soon other bacteria and fungi appear, called mesophilic ("middle-loving") species, which live from about 40-110°F. They continue the decomposition and use the partly broken down intermediate products formed by the cryophilic bacteria as their food, along with the dead bodies of the preceding generations of bacteria. The populations of microorganisms increase at a tremendous rate, as does the temperature.

Next, after several days, the thermophilic ("heat-loving") bacteria take over. They can only live at temperatures above 104°F. Their growth can raise the temperature in the center of a large mass of decomposing organic matter to 160° or 170°F. After several days the temperature will drop and fungi and actinomycete species will break down some of the more resistant components, such as cellulose, hemicellulose, and lignin. The various microorganisms also produce enzymes which continue the decomposition even after they die.

During decomposition the ratio of carbon to nitrogen is reduced from its initial value of perhaps 30:1 or 50:1 to a final ratio of about 10:1 to 12:1. The reason is that the microorganisms use the high-carbon-containing compounds, carbohydrates and fats, for energy, and these materials disappear by being converted to gases: carbon dioxide and water vapor.

In actual field conditions, most organic matter decomposes under both anaerobic and aerobic conditions. On the outside of a mass of organic matter, or near a pore space in the soil, there will be plenty of oxygen, so aerobic organisms will flourish. Inside the organic matter or in poorly aerated soil, anaerobic conditions will exist. Actually, there are

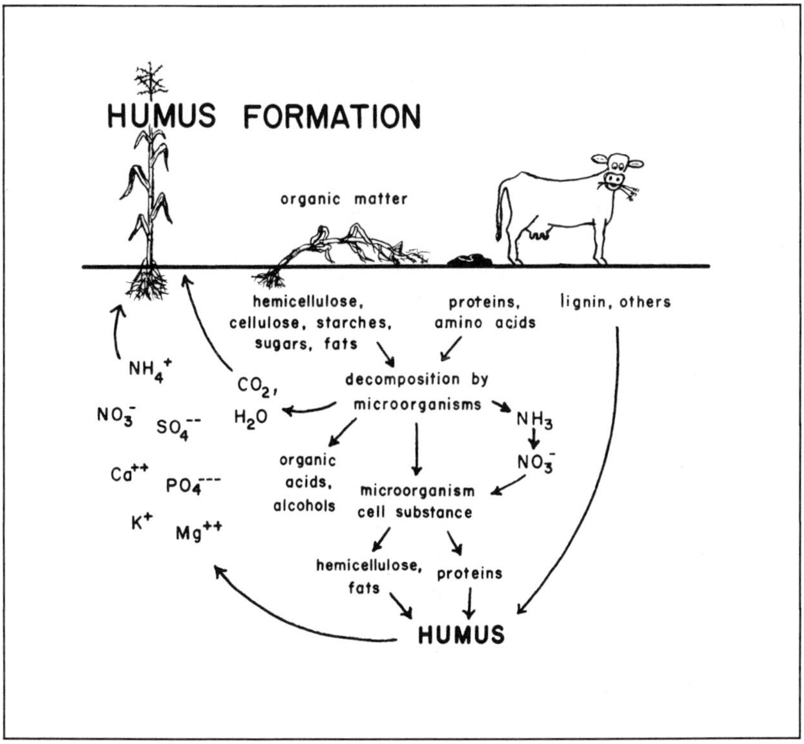

many bacteria which can function as either aerobes or anaerobes, called facultative anaerobes. In general, aerobic decomposition is to be preferred with regard to providing more nutrients to plants and not producing toxic by-products.

Humus. The end product of aerobic decomposition of organic matter is called humus. One ton of cornstalks, leaves, and cobs will produce about 100 pounds of humus. Humus is a complex, colloidal material (composed of small particles) containing proteins, lignin, fats, carbohydrates, and organic acids, as well as other broken-down products of the original organic matter, plus the metabolic products and remains of the microorganisms that did the decomposing.

As we have seen in Chapter 3, humus is a very valuable substance in soil in many ways:

1. It is a storehouse of plant nutrients. Because of its colloidal properties, it can hold three times the amount of nutrients that clay can.

2. It helps make nutrients available to plants. Acids released during decomposition (a) increase breakdown of mineral soil particles (b) may form chelated compounds, which speed the plant's uptake of certain

minerals, and (c) stimulate plant growth. Microorganisms temporarily "tie up" or hold nutrients in their living cells, thus counteracting leaching of the nutrients into the groundwater; the nutrients are later released for plant use when the microbes die. Some microorganisms, especially the symbiotic mycorrhizae, actually "feed" nutrients to plant roots, and of course nitrogen-fixing bacteria supply additional nitrogen. Certain fungi and bacteria protect plants from diseases and pests. The activity of soil microorganisms increases as the temperature rises; thus they provide more nutrients to plants at just the time the plants need them most: the warmest part of the growing season.

3. It improves soil tilth, making soil loose and crumbly. This contributes high water absorption and holding, good aeration, low water and wind erosion, and ease of plowing and cultivating.

Now let's apply these principles of decomposition to the practical operations of farming.

Composting. One direct application of microbial decomposition to agriculture is the practice of composting, which is especially used by followers of "organic" farming and gardening. Composting means letting organic matter rot *before* it is applied to the soil. The resulting product of decomposition is called compost, which is essentially, or nearly humus.

The proponents of composting are generally very enthusiastic about it, even though it requires extra time, equipment, and trouble to do. There are many methods of composting, which we do not need to consider now (see Chapter 13 for details).

Since compost is basically humus, it obviously is very beneficial to the soil. It is especially helpful to well-drained sandy soils, for it reduces drainage (leaching) more than adding fresh, undecomposed organic matter will. Compost has the advantage of being already prepared, "instant" humus. It provides its beneficial properties to the soil immediately, without having to wait several weeks or months for organic matter to decompose in the soil. And then, don't forget that not all soil is "healthy" (well aerated, free of toxins, full of beneficial organisms), so adding fresh organic matter to unhealthy soil will do no good and will probably do harm, because anaerobic decomposition will occur. Compost circumvents this problem, plus helps make the soil healthy and aerobic.

Another problem with adding fresh organic matter directly to the soil is that rapid microbial decomposition will "tie up" most of the nutrients (nitrogen, phosphate, etc.) before the crop plants can use them (as Dr. William A. Albrecht has often written, "The microbes eat first"). In other words, the microorganisms *compete* with the plants for nutrients when their numbers are large, but later when they die, the nutrients are made available to the plants. (This is less true of organic matter derived from legumes than from other sources because legumes contain considerable nitrogen.) For this reason, *large amounts* of fresh organic matter should not be incorporated in the soil just before or during crop growth. Fall application is preferred. But compost doesn't cause this problem and can be applied at any time.

Composting has another plus, if it is done at high temperatures (some composting methods use lower temperatures): it kills weed seeds and disease-causing organisms. But a minus is that some nutrients and carbon dioxide that could be used by crop plants are lost during the composting process. If the soil is "healthy" and fresh organic matter is incorporated in the right way and at the right time, results can be as good as with compost, or even better. So one should weigh the benefits of composting against the disadvantages in extra time and expense.

Animal manures. If done properly, the use of animal manures can be a very valuable asset to soil and crops. If used improperly, manure nutrients can be lost or soil and crops can even be harmed. Manure management will be covered in more detail in Chapter 13. In general, if manure rots aerobically it will help soil and crops by adding organic matter and nutrients. Tests have shown that mixing animal manures with inorganic fertilizers or rock fertilizers gives greatly improved yields.

Green manures & cover crops. The practice of plowing under a growing crop is in general very beneficial to the soil and future crops. Young growth is usually quickly decomposed, adding nutrients. Older plants are more resistant to decay and thus contribute more to the humus content of the soil. If a non-legume is plowed under, rapid decomposition can temporarily tie up nitrogen needed by crops, while legumes can increase the nitrogen content of the soil right away. Thus, a common practice is to plant a mixture of a legume and a grain or grass. Some species that make good green manure include alfalfa, alsike clover, barley, beggarweed, buckwheat, cowpea, domestic and Italian ryegrass, field brome grass, field peas, lespedeza, oats, red clover, rye, soybeans, sudan grass, sweet clover, and vetches.

Planting such temporary crops in the fall (winter cover cropping) provides the additional benefit of protecting the soil from erosion over the winter. A possible disadvantage would be loss of more water from the soil by growing plants than from fallow soil.

Generally, green manures release their nutrients quickly and only benefit one year's crop, and contribute little humus to the soil, while animal manures decompose more slowly, providing nutrients for more than one year, and add more humus. An excellent practice is to incorporate a mixture of the two into the soil, giving the advantages of both. One advantage of green manures is that fresh chlorophyll can absorb toxic substances in the soil. As mentioned earlier, organic matter should not be plowed in too deeply because below several inches, most soils are anaerobic, and proper decomposition will not occur.

Detoxification. A few words about another valuable service provided by soil decomposers: the breakdown, or detoxification, of many of the toxic chemicals that man so liberally (and often unwisely) spreads around the environment.

We are all aware of the extensive use of herbicides and pesticides and that many of them are highly toxic, often to "good" animals and plants, even our crops (herbicide "carryover"). They also may pollute and accumulate in soil, surface waters, and groundwater. Fortunately, many herbicides and pesticides are not quite so dangerous and most do eventually break down, some more rapidly than others. Some become held (adsorbed) by particles of clay and organic matter, where their toxicity is greatly reduced. Some are broken down by sunlight or by chemical reactions in the soil. But a major method of breakdown is by the metabolic activities of living soil organisms. Soil bacteria are especially "talented" at degrading many pesticides, simply by using them as food. High levels of herbicides or pesticides will kill most or all soil life in the local area, but new organisms usually colonize the poisoned area from surrounding soil. Some toxic chemicals are readily broken down (such as the herbicide 2, 4-D) while others are very resistant to microbial attack (the related herbicide 2, 4, 5-T). High soil humus levels (which favor microorganisms) are the most important property facilitating pesticide degradation. Unfortunately, modern farming practices generally reduce or destroy humus, as well as the beneficial soil organisms that produce it, while pouring on more and more poisons to kill the weeds and pests that result from destroying humus and soil organisms. Make sense? No, we're going around in a circle, but it's one of those bad kind—a vicious circle. The kind that close in on you and swallow you.

We have now come to the end of another circle, our journey through the food chain. It was a long trip, but an interesting and exciting one. One that revealed a great deal about how plants, animals, and the soil work. Also one that pointed out some problems. Now let's leave the world of nature and explore the realm of man's society. We may soon find ourselves going around in more circles!

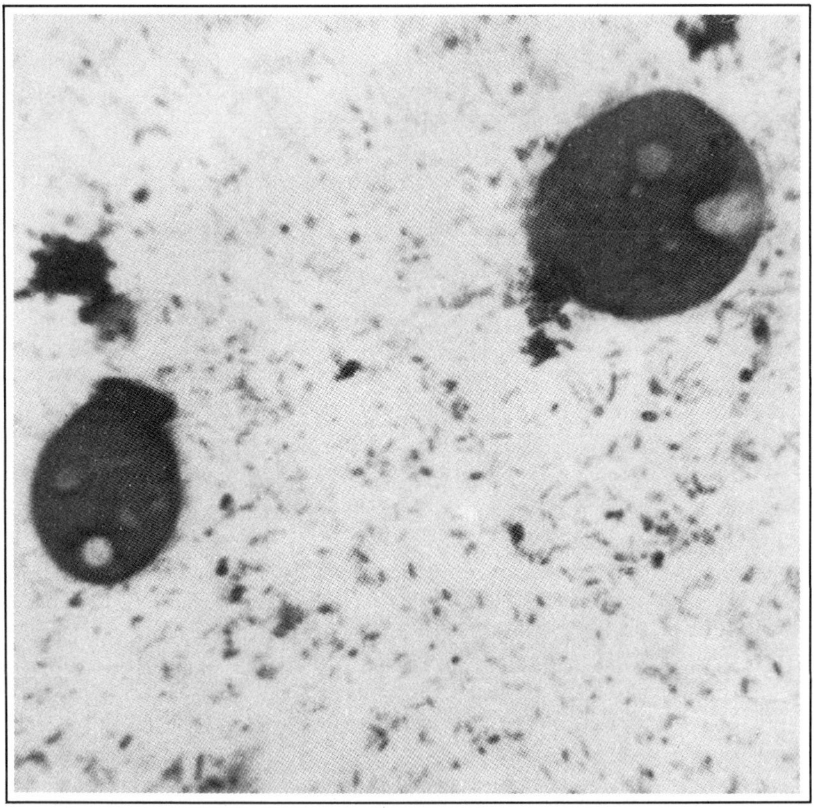

Protozoa and bacteria from decomposing manure.

UNITED
STATES
DEPARTMENT
OF
AGRICULTURE
OFFICES
STABILIZATION

Chapter 7

The Establishment

We have just taken a look at a natural system, the food cycle. It is composed of many parts and "runs" according to definite laws. It runs beautifully if not tampered with too much.

Not let's look at another system—a man-made system. It is also composed of many parts and operates by various laws and rules. It is very complex today, but it gradually developed over the years from simpler beginnings. It may or may not run beautifully, depending on your viewpoint or where you are in it. It is the complex interwoven system of finance-government-industry-business-education, which so greatly affects every aspect of a farmer's life, in ways few realize. And what affects the farmer affects the rest of the country.

Undoubtedly all of the institutions involved would profess to be interested in the farmer's welfare, but don't forget that in our political-economic system, everyone also has allegiances to the group to which he belongs. The businessman wants to make a profit. The politician wants to be re-elected. The bureaucrat wants to justify his job. The professor wants to perpetuate the truth as he sees it. So let's poke around in the "establishment" and see what we find.

Alphabet soup. Just think of all the different organizations and agencies that affect us in this modern age, both directly and indirectly. Besides those that affect every citizen, such as the Internal Revenue Service, U.S. Postal Service, telephone company, local governmental bodies, etc., there are many that affect a farmer's life in one way or another. Let's just list some of the major ones and give their stated functions or interests.

Federal Government.

(Information taken from the *U.S. Government Manual*, 1981/82)

I. *Department of Agriculture*—to improve and maintain farm income and develop and expand markets; to help curb and cure poverty,

The new USDA Dairy Forage Research Center, located near Prairie du Sac, Wisconsin.

hunger, and malnutrition; to enhance the environment and maintain our production capacity by helping landowners protect soil, water, and other natural resources.

A. Farmers Home Administration (FHA)—gives credit to those unable to obtain it at reasonable rates from other sources.
B. Rural Electrification Administration (REA)—finances (by loans) rural electrification and telephones.
C. Agricultural Cooperative Service (ACS)—helps farmers to help themselves through cooperative organizations.
D. Agricultural Marketing Service (AMS)—maintains orderly market conditions for certain commodities, including milk, fruits, & vegetables, by establishing a minimum price and helping stabilize prices.
E. Federal Grain Inspection Service (FGIS)—sets grain standards and uniformity.
F. Office of Transportation (OT)—develops transportation policies with regard to facilities and quality.
G. Food Safety and Quality Service (FSQS)—assures wholesome, safe food and honest labeling.
H. Agricultural Stabilization and Conservation Service (ASCS)—commodity and land use relative to voluntary production adjustment, resource protection, and price, market, and farm in-

come stabilization (through loans, purchases, payments, set aside, acreage allotments, and marketing quotas).

I. Commodity Credit Corporation (CCC)—to stabilize and protect farm income and prices, to assist in maintaining adequate and balanced supplies of agricultural commodities, and to facilitate orderly distribution of commodities (through loans, purchases, and payments).

J. Federal Crop Insurance Corporation (FCIC)—to provide crop insurance for weather, insects, and disease.

K. Foreign Agricultural Service (FAS)—to stimulate overseas markets.

L. Office of International Cooperation and Development (OICD)—coordinates (through land-grant universities) USDA efforts in international development and technical cooperation through research and scientific exchange programs.

M. Soil Conservation Service (SCS)—to develop and carry out a national soil and water conservation program in cooperation with landowners.

N. Economics and Statistics Service (ESS)—economic, statistical, and social research; to collect and publish statistical data.

II. *Department of Commerce*—to promote international trade, economic growth, and technological advancement; to encourage the competetive free enterprise system.

III. *Department of Health and Human Services*—involved with the nation's human concerns, including public health services, food and drugs, social security, youth and the aged, native Americans, and civil rights. Included is:

A. Food and Drug Administration (FDA)—to protect health from impure and unsafe foods, drugs, and cosmetics. Included are:

1. Bureau of Foods—research and development of standards on composition, quality, nutrition, and safety of foods, food additives, colors, and cosmetics.
2. Bureau of Veterinary Medicine—safety and efficacy of veterinary preparations and devices.

IV. *Department of Interior*—the nation's conservation agency, responsible for most nationally owned public lands and natural resources, including wildlife, minerals, and national parks and historical places. Included are:

A. U.S. Fish and Wildlife Service—stewardship of and research on wild birds, endangered species, fisheries, and wildlife; conservation, development, and management of the nation's wildlife resources.

B. Bureau of Reclamation—to protect the environment, develop resources, and stabilize the local economy.

V. *Independent Establishments and Government Corporations*

A. Commodity Futures Trading Commission (CFTC)—regulates trading on the eleven U.S. futures exchanges.

B. Environmental Protection Agency (EPA)—to protect the environment: air, noise, radiation, water, waste management, pesticides, and toxic substances.

C. Export-Import Bank of the United States (Eximbank)—aids in financing; facilitates trade.

D. Farm Credit Administration (FCA)—to supervise and coordinate borrower-owned banks and associations of the Farm Credit System (Federal land banks, Federal intermediate credit banks, banks for cooperatives, and Production Credit Associations)

E. Federal Reserve System—central bank for the U.S.; administers and makes policy for the nation's credit and monetary affairs; to contribute to the strength and vitality of the U.S. economy.

F. Federal Trade Commission (FTC)—to maintain strongly competitive enterprise.

G. Interstate Commerce Commission (ICC)—to regulate interstate surface transportation.

State Government.

State agencies vary in each state, but typically there are at least departments of agriculture and natural resources which have a great affect on farming, plus the state land-grant colleges or universities and their extension system and departments of agriculture.

Private Organizations.

There are literally hundreds of other organizations with a great variety of interests. Some are organizations of farmers, professing to promote farmers' welfare, including the American Agriculture Movement, American Farm Bureau Federation, National Grange, National Farmers Organization, and Farmers' Educational and Cooperative Union of America (National Farmers' Union). Others are business and industry organizations, promoting themselves and their products, such as the Fertilizer Institute, the Potash and Phosphate Institute, Farm and Industrial Equipment Institute, International Silo Association, Irrigation Association, National Feed Ingredients Association, National Hay

Association, etc. (from *Encyclopedia of Associations*, Vol. 1). Such organizations as the Rockefeller Foundation, Ralston Purina Co., National Recreation and Park Association, and the Golf Course Superintendents provide grant money for agricultural research (from *Annual Register of Grant Support*, 1980-81).

Business and Industry.

Then there are all the industries and businesses that manufacture and sell agricultural equipment and supplies, ranging from tractors and plows to milking machines to feed supplements to fertilizers to pesticides and herbicides to drugs and antibiotics. They do a multimillion dollar business yearly, which is not news to any farmer. Other businesses depending heavily on farmers include real estate, supermarkets, publishing, veterinary medicine, soil and fertilizer consulting, and insurance.

Banking and Finance.

The established system of banks and credit agencies have a major involvement with agriculture and farmers.

Markets.

The farmer usually sells his products through the complex marketing system, which includes local markets, national markets, and international markets, along with their attendant brokers and speculators.

Middlemen.

Finally, there are the infamous middlemen who transport, process, store, and distribute the farmer's raw products to the consuming public. Many of them are greatly influenced by labor unions.

Connections. Just the existence of so many organizations and agencies that affect farmers is bewildering enough (and we haven't listed them all), but it is interesting and enlightening to delve into the interconnections and influences of one upon another. Such interrelationships, if they can be discovered, often go a long way toward explaining the loyalties and probable motives behind some of the activities of the organizations.

As we mentioned earlier, some motives are obvious; everyone is interested in promoting the interests of their own group: the farmer wants to prosper, the businessman wants to make a profit, the politician wants to be re-elected, etc. This easily explains why the fertilizer industry and their dealers, for example—or the drug companies and their salesmen—

would promote their products. And the more aggressive salesmen and dealers will make more sales and profits, and rise in the company hierarchy. And the more aggressive *companies* will make more sales and "outcompete" their competitors. That's the capitalistic system we live under—that's the American way.

Unfortunately, human nature being what it is, the "more aggressive" salesmen and companies do not always play fair. Pay-offs, shady deals, "arrangements," quality-cutting, short-weighting, and many other dishonest practices abound. And the examples used here are not isolated or worse than others that could be cited; corruption has now infiltrated every level of society and government.

Fertilizer connections. One illuminating example of interconnections is found in the fertilizer industry. Generally, the farmer encounters this industry when a field representative or salesman from a local CO-OP or independent fertilizer dealer visits his farm, probably around the first of the year, when budgets and crops are being planned. He's not just selling fertilizer, he's selling a *program*. For a certain fee, perhaps $8 to $12 per acre, he promises to (1) test your soil, (2) recommend the right amount of fertilizers (almost always only N, P, K) to raise a certain predetermined yield goal (say, 150 bushel corn, or 50 bushel soybeans), (3) arrange to have the fertilizers applied, (4) recommend the "required" herbicides and insecticides, and possibly (5) perform crop tissue tests during the growing season. He may also offer services in insect and disease scouting, livestock feeds, or even doing your bookkeeping—for an additional fee.

Isn't that great? Sounds like a good deal. You're so busy trying to farm your farm plus several others you've rented, that you would much rather leave those bothersome details to the "experts." Right? Besides, they give away "free" caps or jackets (with their logo boldly emblazoned—ever feel like a walking billboard?). Rare indeed is the farmer who investigates beyond the local dealer. Where does he get his information and advice, anyway? Aren't your soil and crops too valuable to trust to just anyone?

The fertilizer dealer either sends your soil samples in to the state university soil testing lab or to his own company's lab; possibly to an independent lab. They test your soil according to certain standard procedures (see Chapter 10) and return the results to the dealer. Almost certainly, the soil testing lab also had their computer "spit out" fertilizer recommendations along with the test results, so that the local fertilizer dealer will not have to figure them out, since he is probably not an agronomist or soil scientist—his "bag" is selling fertilizers.

University. Well, that's fine, assuming that the soil testing procedures are based on sound principles (Chapter 10 may raise some doubts in your mind about that). But what was the computer's program that figures the fertilizer recommendations based on? The main thing the computerized recommendations are based on is the opinions, "best guesses," and recommendations of the university agronomists and soil experts who helped design the computer program. And those experts' recommendations officially are based on field experiments done by USDA and state university experimental farms. The field experiments are simply "boiled down" and programmed into the computer as a simple multiplier factor; for example, add 1.25 pounds of nitrogen for each bushel of corn desired. This methodology does not take into account the fact that your field and soil may not be like the fields and soil on the experimental farms. It does not consider past history of fertilization and previous crops, nor the extra amount of fertilizer value provided by organic matter and soil organisms.

But an interesting remark made by Dr. William C. Liebhardt of the Rodale Research Center in an AP press release, Jan. 20, 1982, adds another dimension. In a story about unreliable soil testing, Leibhardt said, "The labs are scared to death of nutrient deficiences and hefty recommendations—or over-recommendations—are just built into the system."

Research. Also, a note about the quality of agricultural research in the United States. In a report by Nicholas Wade published in the January 29, 1982, issue of *Science*, a summary is made of a new government report, "An Assessment of the United States Food and Agricultural Research System" (Office of Technology Assessment, Washington, D.C., 1982). The government report states, "Many people, including Congress, have voiced concern that little, if any, overall planning and coordination of research exist, especially at top levels of administration." Wade cites an earlier report, the Pound report, issued by the National Academy of Sciences in 1972, which severely criticized the agricultural research system for its neglect of fundamental biological research and its performance of "a shocking amount of low quality research." Wade says the Office of Management and Budget has been asking why the private sector shouldn't provide more funding for research, rather than increasing government outlays.

Industry. But a solid link already exists between the private sector and agricultural research. For example in the state of Wisconsin, the thick book (over 200 pages) of research reports distributed at the annual Fertilizer, Aglime & Pest Management Conference states on its first page: "Appreciation is expressed to the Wisconsin fertilizer industry for the support provided through the Wisconsin Fertilizer Research Fund for research work conducted by staff members in the Departments of Soil Science and Agronomy. Appreciation for the financial support from the Aglime Association through the tonnage fee for research work on aglime is also gratefully expressed." The tonnage fee is further explained on page 37 of the 1982 report: "The Council . . . approves fertilizer research projects using funds collected semi-annually on each ton of agricultural fertilizer sold in Wisconsin . . . " On pages 36-37 we learn that the Wisconsin Fertilizer Research Council "approved the spending of $134,621 during 1981 on 15 research projects and other activities relating to the correct usage of fertilizer materials" and "set aside $34,000 as a reserve fund to be used for new projects at a later date." The $134,621 divided by 15 projects would average $8,974.73 per project. Not a fantastic amount, but substantial nevertheless. On page 110 we learn that about $240,000 is received annually for research from the sale of cranberries, cherries, potatoes, aglime and fertilizers, accounting for about 7% of total industry funding. And Wisconsin is only one of 50 states.

In the same 1982 book, we notice an article entitled "Mutual Relationships between Industry and the University" by Dr. Leo M. Walsh, former extension soil scientist and now Dean of the College of Agriculture and Life Sciences, University of Wisconsin–Madison. In it we read,

> "The University educates farmers and those who serve farmers through our resident instruction and extension programs. In addition, our scientists conduct research, the results of which are used by the agribusiness industry and are translated into technology which is adopted by the farmer. The University and industry have a common goal; that is, to make agriculture more efficient and to continue to produce ample quantities of high quality food for the Nation and the world.
>
> "We at the University consider the needs and priorities of farmers and the industry when establishing instruction program and research priorities. And the industry has been very supportive of the University in terms of helping to build good public relations and political support with decision-makers. Also, the industry has been generous in terms of providing some financial support for instruction and research programs."

The rest of the article gives details. Comparing Wisconsin agricultural research funding for 1970 and 1980, we find that state funding has fallen from 51.9% in 1970 to 39.2% in 1980, while industry support has stayed at about 10% (10.8% in 1970, 9.7% in 1980). The balance was made up of federal support, both formula funds (Hatch) and competitive grants. The actual dollar amounts in research budgets grew from about $10 million in 1970 to $30 million in 1980, so the 10% industry support has grown from about $1 million to $3 million. Industry funds contribute to university instruction partly through scholarships ($118,826.95 in 1980-81), an internship program (placing about 325-350 students annually in agricultural businesses and agencies), a minority students program (scholarships, tutoring, recruiting, and internships), and career advising and placement services.

In a separate section on University Extension, Dr. Walsh notes that "Industry has worked closely with extension specialists and county agricultural agents for many, many years." Some of the agricultural programs and activities of Extension are listed: (1) credit courses and short courses (including short courses for dealers and salesmen), (2) "in depth" clinics for farmers and dealers, (3) specialized meetings or workshops for industrial representatives, (4) experimental farm field days and county-based demonstration plots, (5) use of mass media, (6) use of written material and visual aids, (7) use of computers to aid farmers, agri-business representatives, and others in decision-making, and (8) specialized projects, such as soil test summaries, plant analysis summaries, and projections of fertilizer use.

Finally, in a summary, Dr. Walsh states . . . "I think we can be proud of the University-industry relationships we have developed in the past. But, this is no time to rest on our laurels. We need to strengthen these relationships . . . " Besides the obviously close university-industry relationships noted in Dr. Walsh's article, did you notice his reference to two others: "political support with decision-makers" and the "use of mass media"? The activities of political lobbyists are well known and will not be elaborated on here. The effectiveness of the mass media in shaping and directing public opinion is also well known.

Some interesting observations can also be made about the University of Wisconsin-Extension publications, available in every county extension office. There are hundreds, ranging in subject from soils to livestock housing to insect pests to home economics. Most are written and illustrated in a rather sedate, dry academic style and contain much valuable information. One in particular stands out: publication number A2240, "Aglime, Key to Increased Yields and Profits" (1970). It can be described in no other terms than a real selling job, full of slogans, pictures, tables, and graphs (and very little text) showing the benefits of using aglime (dolomite). Dolomite contains 22% calcium and 14% magnesium. Most of Wisconsin's underlying limestone is dolomite, so most soils have plenty of magnesium (this is not true in many other states). Crops need more calcium than magnesium, and if they take up too much magnesium they are of poor quality. In Wisconsin, other sources of lime or calcium should be used instead of dolomite. In at least four other extension publications, numbers A2523 (1973), G2986 (1979), A1880 (1974), and A2448 (1977), aglime is recommended. In fact, in G2986, non-magnesium sources of calcium are not recommended for fear of causing a magnesium deficiency. Magnesium also causes soil nitrogen to be lost into the air, thus requiring the addition of more nitrogen fertilizer than if non-magnesium lime is used.

Questions. So what have we found in following connecting threads? We found that those who do most of the soil testing and fertilizer recommending are the dealers who sell fertilizer and that the recommendations are based on university research partly bankrolled by the fertilizer industry. Might these facts influence the fertilizers and amounts recommended? Or even the results and conclusions of the research?

Others. There are other interesting connections, which we cannot take the time to explore in detail. Could there be any outside influences on the treatments and remedies recommended by veterinarians, considering that they are educated at state universities and that drug companies have a "vested interest" in promoting sales of their products?

And why do university professors, informational pamphlets, and extension agents consistently recommend toxic chemical herbicides and insecticides to "control" weeds and pests, rather than recommending methods of non-toxic *prevention* by working *with* nature, not against it. Much is already known and published about non-toxic weed and pest control, available to the professors if they would be willing to consider it.

A great deal of blame can be laid on the modern educational system, which perpetuates a body of "knowledge," part correct, but part false (as revealed by new discoveries made almost daily). The ivory tower professors and the textbooks they write are sometimes as slow as a glacier in considering and accepting facts that contradict their beliefs. Their minds have been so thoroughly indoctrinated by the knowledge they received from the previous generation of teachers, that some "experts" simply cannot conceive that there could be other ways of thinking. Worse yet, they or the technicians and salesmen influenced by them may even attack or sabotage others who promote a contradictory set of beliefs. Such is human nature.

Markets. And how are the market prices which farmers receive for their commodities determined? Do the markets really operate on supply and demand? We all know that prices are in part controlled by the government's economic policy, or farm policy. And who sets economic policy? Economists, politicians, bureaucrats, and advisors (who probably have never set foot on a farm) play a large role, especially the Committee for Economic Development, a prestigious group of 200 leading businessmen and scholars. Also the President's Council of Economic Advisers, composed of three members appointed by the President and approved by the Senate; and the Council on Wage and Price Stability, which monitors the economy as a whole, play an important part, along with the congressional Joint Economic Committee of Congress. Then there is the Federal Reserve System, directed by a 7-member Board of Governors, appointed by the President; and by a 9-member Board of Directors in each of the 12 districts (3 members being bankers elected by commercial banks in the district, 3 are businessmen elected by banks, and 3 are residents appointed by the Board of Governors). Are these people really sensitive to the needs of farmers?

In a 1962 recommendation of the Committee for Economic Development, "An Adaptive Program for Agriculture," this group of 200 presidents and board chairmen of large corporations spelled out a long-range program to move people and land out of agriculture. They said, "Excess resources in use in the production of farm goods *is* the farm problem." On the previous page, they explain excess resources as mainly labor and possibly capital.

How would this be done? ". . . 44 per cent of the farm population is below the age of 20. Here in our opinion, is a main key to agricultural adjustment: we have an opportunity to secure a long-lasting relief from the overburden of people pressing upon farm income by getting a large number of people out of agriculture before they are committed to it as a career."

Also, "the maintenance of employment opportunities in nonagricultural industry and services is an essential condition for the most satisfactory agricultural adjustment. . . . resources, particularly labor, are engaged in farming when they could produce more, and earn more, outside agriculture."

Third, "The basic adjustment required to solve the farm problem, adjustment of the resources used to produce farm goods, cannot be expected to take place unless the price system is permitted to signal farmers how much is wanted, of what. *Therefore, it is recommended that a Price Adjustment Program be instituted* . . . . it is recommended that the price supports be reduced immediately . . . the lower price levels would discourage further commitment of new productive resources . . ."

Shocking? Sound like what's happening these days? Read on. . . .

CHAPTER 8

Why Problems?

Have you ever thought—on one of those days when both the tractor and manure spreader broke down, 5 cows are sick with milk fever, the land rental bill is overdue, and it hasn't rained for a month—that everything is against the farmer? . . . . . and that's one of your *good* days!

Surely the vast majority of farmers would not be doing what they are doing if they did not have a real love for the land, for raising crops and animals, for the (relatively) good clean lifestyle provided by rural living.

PROBLEMS! Everywhere you look, there are problems. The weeds and bugs are worse than they used to be. More cows (or pigs or chickens) are sick or dying than ever before. Expenses and taxes are higher than ever. Bank credit is tighter. More government regulations. Even the soil is getting harder and harder to plow. It's not just your imagination. Read some recent newspaper and magazine headlines: "Agriculture Budget Cut." "Weeds are Farmers's No. 1 Threat to Crops." "Grim Year for Dairymen." "Grain Prices Slide." "Trade Sanctions Hurting U.S. Farmers, WFU Board Charges." "Record Crop a Mixed Blessing." "Plans to Foreclose FmHA Loans Denied." "Serious Problems Seen for Debt-ridden Farms." "Acreage Cuts for Grain Ordered." "Outlook for Farms Gets Worse."

Or consider these facts:

1. An average of 2000 farms have gone out of business *every week* since 1950. Farmers comprised 30% of the population in 1920, 16.6% in 1950, and about 3% in 1980.

2. Farmers' income *derived from farming* (not supplemented from outside employment) has fallen from 5.7% of the national income in 1950 to 1.7% in 1979, while production has steadily risen.

3. Farm taxes (property, interest, and mortgage) have grown about 5 times larger in the last 20 years, and over 14 times larger in the last 40 years.

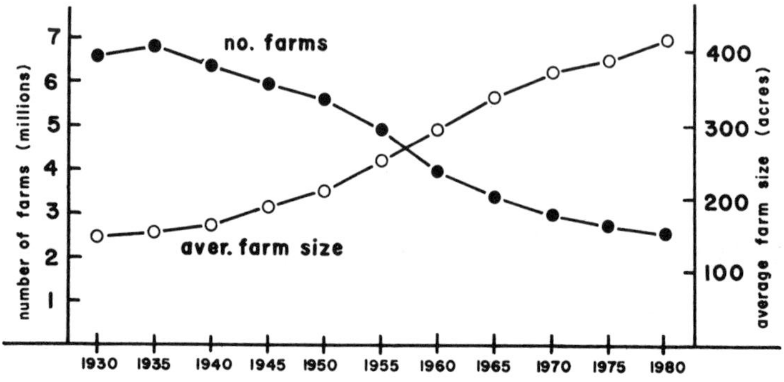

The trend over the last 50 years has been for the number of farms to drop drastically, but for the size of farms to increase.

4. The farmer receives about 38 cents of the consumer's food dollar. About 24 cents goes for processing, 10 cents for wholesaling, 22 cents for retailing, and 5 cents for transportation.

Even rural living is no longer what it used to be. Crime and drugs have invaded the country. In an article, "Farmlife can be Hazardous to Your Health," in the December, 1981, issue of *Successful Farming*, Dr. Kelly J. Donham cited farmstead health dangers of mental stress, lack of exercise, farm machinery accidents, toxic agricultural chemicals, cancer, and toxic gases in grain and liquid manure storage structures and livestock confinement buildings.

Why all these problems? And why are they getting worse and worse? As we have said in Chapters 1 and 2, we should first carefully analyze problems, note their symptoms, and determine their causes before attempting to solve them. And the only permanent solutions can come from attacking the causes, not the symptoms; and once the causes are eliminated, preventing them from recurring.

Factors. In the complicated business of growing plants and animals, there are many factors and influences that can cause problems. There is the weather, which is pretty much uncontrollable by man (although there is evidence that man's activities do influence weather). Droughts, flooding, hail, and windstorms can destroy crops. Then there are soil, plant, and animal nutrition. These factors are most certainly under the farmer's control. The genetic strains and breeding of crops and livestock can be chosen at will. Tillage and planting methods have a great deal to do with production, as do animal husbandry and other general management methods. The farmer can choose his time of planting, cultivation, harvesting, and his use of herbicides and pesticides, plus which animals to buy, sell, and breed. He can hire extra help, rent more land, build more barns and silos, and buy whatever machinery and equipment he wants (as long as he can get credit). He can usually market his products when and where he chooses. However, man-made laws and agencies do impose some restrictions on activities involving water, soil conservation, manure disposal, acres tilled, money available, and desirable times to market products. Certainly taxes impose a considerable burden.

In control? But to a very great extent, the farmer can be and should be in control of his situation and well being. But is he really? Or does he let others tell him what fertilizers to use and how much, which herbicides and insecticides to spray and when, where to sell his life-giving products and what price he will receive? Advertising by business and industry plays a tremendous part in determining what brand of seed corn, herbicides, tractors, milking machines, and silos you buy. And even our neighbors influence us, for few farmers want to do anything much different from their neighbors, although most do want to have a newer and bigger tractor and silo. Unless your farm is paid for and you have been very lucky in weather and yields, your banker or loan officer probably

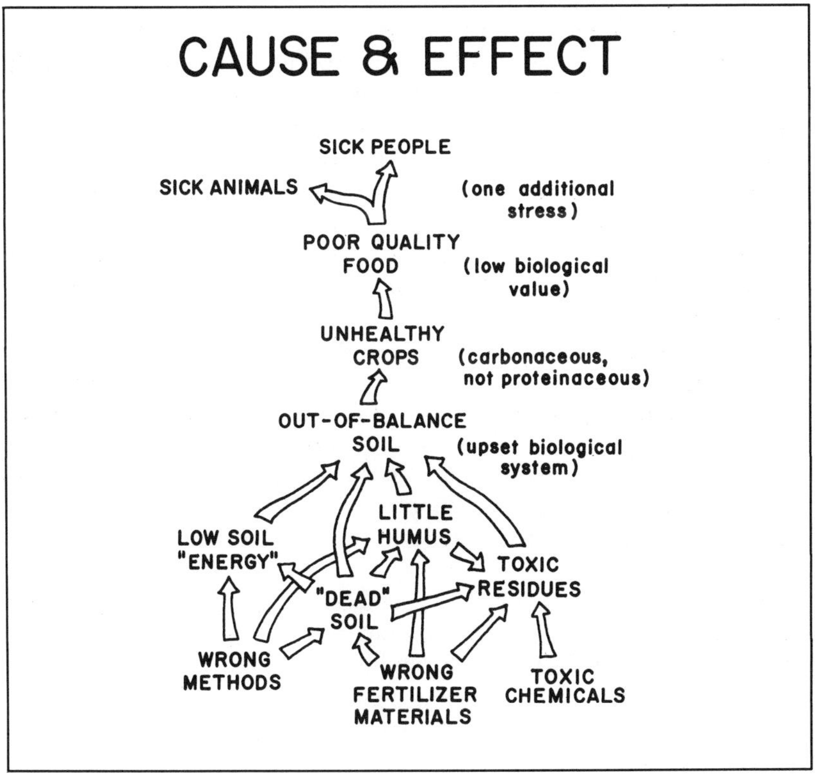

plays a great role in literally running your farm. We tend to let the "experts" tell us how to till and fertilize our soil, treat our sick animals, and sell our produce. Is bigger always better? Are we devoutly bowing down in the temple of big agriculture to the high priests of technology, before the idols of 300 horsepower dual wheel tractors and tall silos? Are we helplessly caught in the middle of a spiderweb of tangled threads?

Besides the problems that unfavorable weather can create, are many if not most of our problems created, or worsened, by ourselves—by both farmers and those who advise or regulate them? Let's look at some possibilities.

Nature's problems. If you have read through Chapters 3-6, you should realize the vital importance of fertile "living" soil in producing high quality crops and healthy animals, for the soil is the absolute basis of agriculture. Anything the farmer might do that will harm his soil and soil life in any way will cause problems. And any other bad management practices involving crop and animal raising will add to those problems. Recognizing and solving these problems is what this book is mainly about, so more will be said in later chapters.

Man's problems. Would a university professor or extension agent or fertilizer dealer recommend a fertilizer program to you that will actually (1) kill beneficial soil organisms, (2) make your soil hard and more easily eroded, (3) destroy humus, (4) cause nitrogen to be released into the air and leach into the groundwater, (5) grow sick, poor quality crops, and (6) promote weeds, pests, and plant diseases? Impossible, you say? Would you believe they've been doing it for decades? As we shall see in Chapter 12, such commonly used fertilizer materials as anhydrous ammonia, muriate of potash (0-0-60 or 0-0-62), and dolomite ("aglime"), plus over-use of many salt fertilizers, can do just that.

But why would fertilizers be recommended that harm the soil? One reason could be that by killing soil organisms and throwing soil chemistry out of balance, conditions are created which favor weeds and pests—and require more chemicals to control them, plus requiring more and more fertilizers to grow the same yields of crops. And what salesman would refuse larger sales of his products? And don't forget how we learned in the preceding chapter that the university is partly funded by industry.

Can you believe that hybrid seed companies "design" their hybrid varieties for various types of "sick" and out of balance soil? Hybrid varieties are recommended not only on the basis of climate and desired maturity time, but also for high potassium or high magnesium soils.

Have you ever noticed how government programs and regulations so often seek to *control* us? Looking at the functions of the federal agencies listed in Chapter 7, we notice the words, "protect," "promote," "stabilize," "regulate," "facilitate," "supervise," "coordinate," "makes policy," "maintains orderly market conditions." And then, it seems as if the banking industry and the marketing system manipulate prices, interest rates, and credit to their own benefit. Is this just a persecuted figment of our imagination, or is there something wrong with our economic system? How many farmers have to buy the tractors and silos, or plant the number of acres, that their bankers allow? Is this the land of the free, or not?

In a revealing study of the U.S. economic system (*The National Economy is out of Balance*, 1966), Dr. John Forbes proved by using government statistics that since about 1950, the economy has been growing increasingly unbalanced with a deceptive "prosperity" being fueled by unjustifiably high wage income and excessive private debt and interest. At the same time, farm income has gotten progressively smaller, proportionately, after farm prices were taken off of 90% of parity and the "sliding scale parity" (90-60%) was put into effect. From the 1946-1950 period, when the economy *was* in balance, until 1964, the gross national product grew 149.5%, wages increased 164.9%, net private debt bulged 306.5%, net interest ballooned 523.3%, while farm income *decreased* 15.9% (Forbes, p. 44-45).

Dr. Forbes, a college history professor, was stimulated to make his study after hearing Carl H. Wilken speak out on the injustices to farmers. Wilken was an economic research analyst and auditor in the National Foundation for Economic Stability, Washington, D.C., as well as a farmer for 20 years. His story is told in the book, *Unforgiven*, by Charles Walters, Jr. (1971). Wilken's research into government economic statistics revealed what appears to be a constant relationship: that the nation's annual earned income is always about in the same ratio as the raw materials income (from agriculture [70%], petroleum and mining, fisheries, and lumbering). Thus, underpaying especially the farmer decreases the whole nation's income.

Both Dr. Forbes and Carl Wilken point out some of the man-made problems in our national economy and farm program:

1. Since 1950, farm income has not grown *in proportion* to the other sectors of the national economy.

2. This is because farmers have been underpaid for their commodities; that is, market prices have been too low.

3. Farmers are the only major producers in the nation who set no price on their products, but who accept whatever price the market gives

them. Farmers as a whole are unorganized and sell as individuals at wholesale, but have to buy supplies and equipment at retail.

4. Market prices have been deliberately held low to cause farmers to reduce production because of supposed "surpluses," which we are told are the cause of the problem. Government subsidies and production quotas are used to regulate production and marketing. Thus, the market is not free to operate on the law of supply and demand. Export of "surplus" farm products is being promoted to create a "favorable balance of payments" in international trade.

5. But "surpluses" do not really exist. Domestic production has not risen as fast as the population. Carl Wilken cited data from the 1944-1945 *Statistical Abstract of the United States* (p. 56) that in the 11 year period between 1933 and 1943, the U.S. imported more farm products than it exported, $12,786,725,000 worth of imports compared to $8,723,787,000 worth of exports.

The items imported do not consist only of products not grown in the U.S., such as coffee, tea, spices, and rubber, but imports of beef and dairy products have been steadily increasing, even though domestic production could easily supply the need. In 1962, 55% of agricultural imports were partly competitive with U.S. products. (*Farmer's World*, Yearbook of Agriculture 1964, p. 370). Compare these figures from the 1980 *Statistical Abstract of the United States*, p. 711:

Beef, in millions of pounds:	1960	1965	1970	1975	1979
Production	14,728	18,699	21,651	23,976	21,400
Exports	55	91	101	110	215
Imports	760	923	1,792	1,758	2,405

However, U.S. export figures after 1942 include food for foreign relief and charity ("giveaways"), which does not increase dollar trade. From 1954-1979, donations and loans made under the Food for Peace program totaled $28.86 billion worth of food (p. 712, 1980 *Statistical Abstract of the United States*).

Part of the "surplus problem" is domestic underconsumption of certain items, related to the post-World War II shift in diet away from a wholesome grain-vegetable-fruit diet towards an overprocessed "junk food" diet, high in sugar, fats, and meat. Compare these per capita consumption figures from the 1960 and 1980 *Statistical Abstract of the United States* (p. 85 & 131):

Item (pounds per capita)	*1910*	*1920*	*1930*	*1940*	*1950*	*1960*	*1970*	*1979*
meat (excluding fish and poultry)	146.4	136.0	129.0	142.4	144.6	173.6	192.6	181.3
fats and oils	N.A.*	N.A.	N.A.	46.4	45.9	45.3	53.0	57.2
sugar & corn syrup	81.9	96.1	122.8	106.5	114.5	111.2	122.6	140.5
fruit, fresh	137.9	145.4	133.6	142.1	107.4	93.9	81.4	84.6
potatoes, fresh	198	140	132	123	106	108	118	123
potatoes, frozen	N.A.*	N.A.	N.A.	N.A.	N.A.	2.7	11.1	18.1
fluid milk and cream	315	348	337	331	349	322	264	231

* not available

Dr. Harrell De Graff, Cornell agricultural economist, has said that a slight increase in livestock could eat up all the so-called surpluses. Yet we are importing an increasing amount of meat for domestic consumption. "So in addition to cutting beef production, imports displace the consumption of our feed grains also. This results in a deliberately created feed grain surplus" (from a speech by Arnold Paulson, Economic Seminar, Feb. 2, 1971). In his book, *Ill Fares the Land* (1966), Dan Van Gorder shockingly states that the U.S. has not been self-sufficient in agriculture since the mid-1920s, when farms first did not produce enough to meet domestic plus export demands (p. 13), and that *per capita* production of most farm commodities reached a peak in the late 1880s and first decade of the 1900s (p. 44). Production has not kept pace with population growth (p. 17-19). According to the Mar. 15, 1966, issue of the U. S. Department of Agriculture publication, *Foreign Agriculture*, "The United States is the world's second largest agricultural importer."

In his comprehensive book, *Prosperity Unlimited* (1947), Carl Wilken stated that surpluses are needed for a prosperous economy, and they should not be sold at a lower price because they are surplus. "Without a surplus we would have nothing to trade for goods which we do not produce, nor would we have the materials with which to expand our economy so as to keep pace with population growth and technological improvement" (p. 8). "Scarcity of production is an economic vacuum which creates neither jobs nor income" (p. 102). "What to do with surplus production is a question that can be answered quite simply by the statement, 'Use it,' and thereby increase our standard of living" (p. 106). "Production creates its own demand through the income from production, processing and distribution" (p. 131).

6. An economy financed by debt is not sound. Excess growth of credit and debt is the main cause of inflation. When borrowed money is spent in the economy, it does generate income, but does not produce new wealth, and the debt is still unpaid. New wealth and the new income

The Case of Casein

Imitation cheese containing casein and/or non-dairy (vegetable) ingredients constituted about 5% (200 million pounds) of U.S. cheese production in 1982 (up from 2% in 1978). But casein, which is a protein milk by-product, is not produced in the U.S., but is imported, mainly from New Zealand, Ireland, and Australia (*Hoard's Dairyman*, Jan. 25, 1982, p. 116-117.

Over one-third of pizza sold in retail stores contains imitation cheese and is priced an average of 21% lower than pizza containing only natural cheese. Casein imports have doubled since 1965, and increased about one-third during the 1970's. In 1955, 99% of casein was used for non-food industrial products. In 1980, 87% of casein was used in food and animal feed (Staff Paper Series No. 205, by T. F. Graf, Univ. of Wisconsin-Extension, Feb. 1982, p. 4-5).

U.S. agricultural exports have helped keep recent U.S. balance of trade deficits from being as disastrous as they might otherwise have been. In fiscal 1980, agricultural exports brought in $40 billion while imports cost $17 billion, giving a surplus of $23 billion to help pay for importing other goods (1981 Yearbook of Agriculture, *Will There be Enough Food*?, p. 216). But within agricultural trade, dairy imports were *higher* than exports.

U.S. dairy manufacturers cannot compete with many dairy imports, including casein, because some foreign countries sell dairy products at low world prices (1/4 to less than 1/2 U.S. domestic prices), even though they give their farmers high subsidies (T.F. Graf, U. Wisconsin-Extension Staff Paper Series, No. 205, Feb. 1982, p. 1-3, 7).

The Multi-Lateral Trade Agreement of 1979, resulting from international trade negotiations in Geneva, Switzerland, went into effect in January 1980 as the 1979 Trade Act. In it, we gave import concessions including reducing agricultural tariffs by over 50%, including a tariff reduction on casein of 1.3 cents per pound. At present there is no import quota on casein. Import quotas on cheese were increased. These concessions were made at the request of the EEC (European Economic Community), which has a 15% dairy surplus. These dairy concessions cost the U.S. $316 million annually, while other agricultural sectors were aided by the Geneva agreement ("Impact of the Trade Act of 1979 on the Dairy Industry," by T.F. Graf, U. Wisconsin-Extension, Jan. 1980, p. 2-3).

These factors have potential severe negative effects on the U.S. dairy industry, which is suffering from political attack because of a slight (1.6% in 1980) surplus in production. This surplus could easily be eliminated by increased domestic consumption (which has been falling since the 1950's) and by feeding calves milk instead of "milk replacer."

it generates in the economy can only come from the production of raw materials. This would include the mining of ore and drilling for gas and oil, but those are non-renewable resources. Only raw material production involving the growth and reproduction of living organisms can produce an inexhaustible source of raw materials. This would include fishing and forestry, but especially agriculture. A farmer can put a grain of corn in the ground and in 80-120 days have at least one ear of corn with 600-800 kernels identical to the one planted.

7. The gross income earned by farmers for their raw commodities, when spent by them, generates about $7 of income in the economy for each dollar spent (this 7:1 ratio has held true since 1921; before that, 1 farm dollar generated less national income). This happens by a pyramid effect as the farmer buys equipment and supplies and stimulates the business of rural communities and the industries that make and distribute consumer goods, plus including the income generated as processors and middlemen market the farm commodities.

The income generated by the farm dollar is greater than that of any other segment of society. Compared to the 7:1 farm ratio, one dollar of wage income only generates about $1.56 of national income. For each dollar of new capital debt, about $4 of national income is generated, BUT it takes about $5.50 of national income to generate a dollar of profit with which to pay interest or liquidate capital debt. From 1929 through 1961, corporations increased their net worth 140.9% but their net debt increased 252%.

8. When farm commodity prices are held lower than a full parity level (in proportion with wages and retail prices of the rest of the economy), the national income is deprived of about seven times that amount, leading to a reduction in all the steps of the economy affected by agriculture. When farm prices were at or near parity level (before 1950), the economy was in balance, the national income grew but debt and inflation did not grow disproportionately. The nation was prosperous. When wages are allowed to grow disproportionately large and raw material prices are kept low, not enough national income is generated to pay for the increased wages, so borrowing must occur to make up the difference. This debt and its accompanying interest load, plus the reduced farm income, depress the economy, since consumer purchasing power is down. This leads to surpluses in the stores and on the farm, not surpluses from overproduction but from underconsumption. This leads to unemployment, more government money (from taxpayers and money borrowed against the future) spent on welfare and unemployment, and on farm price subsidies and storage facilities for the "surpluses." Taxes are raised to help pay for it all, and around we go! As Dr. Forbes put it: "*The*

United States economy since 1950 has been the economy of a depression hid by debt."

9. Rather than blaming the farmer for "overproduction," we should look in the direction of the economists and advisers who set economic policy. We have been operating the economy under the theories sometimes called the New Economics, which are an outgrowth of the older Keyncsian economics put forth by John Maynard Keynes in the 1920s. These economic theories state that capital should be devoted to (1) trade, (2) manufacturing, and (3) agriculture and raw materials production, in that order. That adequate farm prices are a hazard to earnings, but high wages are not. Cheap farm prices (cheap food) give people more income to spend on factory output. Dan Van Gorder in *Ill Fares the Land*, summarizes the Keynesian theory as "spend yourself out of debt." Keynes did not worry about inflation at all in his theory.

Another aspect of the present economic and farm policy is the promotion of increased world trade, and "free trade," with a reduction or elimination of protective tariffs. That way our supposed agricultural

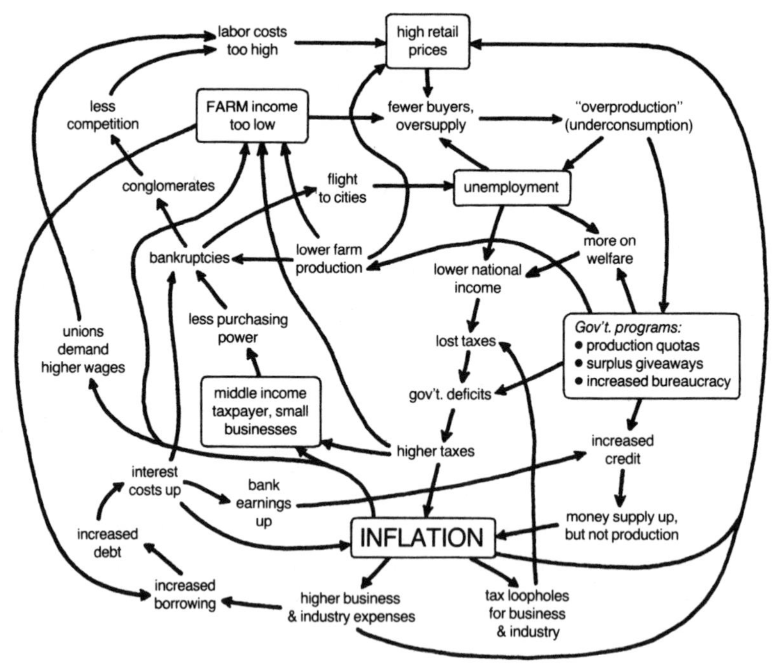

"Oh, what a tangled web we weave..."

"surpluses" can be shipped overseas to help keep the balance of payments more favorable (we have nevertheless been in the red in recent years) so we can buy more imports. That's fine if we import things we really need, such as oil (but do we really need so much of it?), but because our domestic food needs are not being met by restricted production quotas, we are also importing food unnecessarily.

Why are such illogical and obviously harmful and unworking economic policies being followed? Why are farmers most of all apparently being exploited and manipulated? In *Ill Fares the Land*, Dan Van Gorder states that the Keynesian economic theories promote socialism, and he traces them back to Fabian socialism of the late 1800s. This brand of socialism states that socialism will overtake the world *gradually*, rather than by revolution as in Marxism. Van Gorder believes that American agriculture was started on its downward path beginning with the New Deal programs and regulations of 1933. It was at this time that the "overproduction hoax" was originated to justify getting control over agriculture and to prevent farm bankruptcies during the Depression. Van Gorder even shows that the official government statistics were "revised" in 1933 to support the "overproduction hoax," and Forbes also found that statistics going back to 1929 were changed without explanation in the 1966 "Economic Report of the President" compared to the 1965 "Economic Report of the President." Such tactics are apparently still in vogue; a syndicated newspaper article that appeared April 13, 1982, says that "the Reagan administration is using two sets of figures to estimate the 1983 budget deficit, depending on whether officials are talking privately to a roomful of powerful congressmen or to the general public."

Van Gorder says that the Keynesian philosophy is to socialize a country gradually, to allow private ownership but have government control. He cites general collectivist/socialist tactics of controlling the farmers first because being more independent, they are harder to socialize. He believes there is a concerted effort to force farmers off the farms and into the cities and factories, and to subsidize those remaining to produce less. Eventually, through higher prices, taxes, and government controls, free enterprise will be destroyed, and everyone will be made a ward of the federal government.

In another shocking book, *The Naked Capitalist* (1970), W. Cleon Skousen believes that the real motive force behind communist and socialist movements is a network of intellectuals, the few dynastic families of the super-rich, and the international bankers, who have great monetary power and close connections with governments. His book is based on another book, *Tragedy and Hope—A History of the World in Our Time* by Dr. Carroll Quigley, a professor who was an "insider" in this elite group of intellectuals who, with good intentions, want to "save" the world and bring about a man-made millennium of peace and prosperity.

Whether or not you want to believe in the existence of a worldwide or a national conspiracy, you must admit that many organizations and policies in the last several decades *have* brought more and more regulation and control upon us, and the farmer hasn't exactly fared well.

Solutions. These man-created problems are so great and daily getting worse, that one might like to throw up one's hands, say "What's the use?", and give up farming. Many have already. Don't give up hope. There are solutions.

In their books, mentioned earlier, Dr. Forbes, Carl Wilken, Charles Walters, Jr., and Dan Van Gorder give some steps which could go a long way toward correcting the problems:

1. Restore farm prices to a parity level and bring the economy into balance, with the different sectors growing proportionately. Eliminate subsidies. A full parity program will not cost money since farm income generates about $7 for each $1 spent. Parity prices don't lead consumer prices, they follow them.

2. Establish a farmer-controlled food processing and distribution system.

3. Develop local markets in cooperation with business and civic organizations.

4. Maintain protective tariffs at the parity level, representing the difference between the American price level and the price of imported products at the point of entry. When the world price reaches our level, all tariffs could then be at zero, or on a free trade basis.

5. Diversify crops in an area; suit them to soil and climate. We could even grow coffee, tea, spices, and rubber in our southern states and become nearly self-sufficient.

6. Use farmers as policy advisers; have farm referenda before making laws affecting farmers.

7. Strengthen family and youth activities to keep youth on the farm.

8. Return 15-18 million urban people to the land (those with previous farm backgrounds).

9. Break away from reliance on herbicides, pesticides, and commercial salt fertilizers and re-establish ecologically sound agriculture.

10. Educate people to the fact that farming is our most important industry.

A real solution to the farm problem, and many others, cannot come until those who are in charge, who are making policy and directing programs, who are working in manufacturing and business, are not motivated by such human failings as greed and lust for power and self-importance.

Ecology Again

In his book, *The Closing Circle* (1971), Dr. Barry Commoner cuts through the complexities of ecology and the problems of pollution, and exposes the causes of problems. He simplifies the principles of ecology into four laws:

1. "*Everything is connected to everything else.*" That is, all parts of the ecosystem are interrelated, and a change in one part will affect others.
2. "*Everything must go somewhere.*" Matter is not destroyed; it is merely transferred from place to place, converted from one molecular compound into another, and acts on the life processes of any organism it becomes part of.
3. "*Nature knows best.*" Any major man-made change in a natural system is likely to be detrimental.
4. "*There is no such thing as a free lunch.*" Every benefit is gained at some cost because of the interconnected ecosystem (law no. 1); payment of the cost cannot be avoided, only delayed.

Dr. Commoner traces the blame for our present ecological troubles first to the great increase in technology and production of pollutants after World War II (p. 140, 144), and second to a failure to recognize the importance of nature's complex systems and to see the overall effects of our actions (p. 189). But he also goes a step further toward the ultimate cause in pointing out the motive of profit-seeking (p. 258).

Your part. We have seen some mind-boggling facts and ideas. You may feel pretty helpless to change the national farm policy and economic system right now today—so what CAN you do to help solve problems. As an individual farmer, you can certainly change those things you *do* have control over: your own farming methods—your fertilizing and tillage methods, crops grown, animal management, machinery owned, and even marketing. The remaining chapters will show you how.

REEDSBURG

CHAPTER 9

Changing Course

Are you fed up with things the way they are? With decreasing crop quality, sick animals, weeds, bugs, higher fertilizer and veterinary bills and lower market prices, tighter bank credit? Yes, but are you REALLY fed up? Enough to actually DO something about it?

You don't have to be a slave to the traditional system of agriculture, with its over-emphasis of *technology* and *quantity*, and its neglect of *natural systems* and *quality*. You *can* break away. You *can* change.

Wouldn't it be great to have loose, crumbly soil that didn't wash away in an average thunderstorm? To stop having to poison your land (and yourself) with insecticides and herbicides? To need *less* fertilizer each year, not *more?* To grow such high quality crops that you can actually feed *less* feed to your animals and get *more* production, and have few if any sick animals? To be able to rent *less* land and grow *more*? To be able to market your produce where you want, when you want, and at *your* prices? It's not a dream. *Farmers are doing it now!* And they don't necessarily have rich, loamy soil to start with, either—sometimes just "blowsand."

If you really do want to change your farming methods—or even just seriously investigate the possibilities—what do you have to do?

Is it right? The first thing is to become fully convinced in your mind that natural systems farming is the right way to go. Does it make any sense to use dangerously toxic insecticides on crops that people or animals are going to eat? Do you like being locked into a certain crop rotation because herbicide carryover is so great? Does it make sense to use fertilizers that sterilize the soil of its beneficial organisms and destroy vital humus? To grow crops that are so sick and lacking in nutrients that not only do they succumb to pests and diseases, but they contribute to the ill health of animals and humans?

Does it pay? Isn't it more desirable to nurture and care for what provides you and ultimately all Americans their living—your soil? Yes, you say, but does it pay? Can I make it financially by farming the natural systems way? Absolutely. Farmers are doing it now. Who can argue with higher yields and better quality (premium market prices) *plus* lower herbicide and pesticide, veterinary and doctor bills? It is true that during the transition period when you change over from your present methods to natural system methods, you will have to delicately balance purchases of new fertilizers and equipment (if needed) against lowered chemical and medical bills. But encouragingly positive *results* will be seen right away the first year.

Scientific studies comparing conventional farming methods with "organic" and Amish farming methods (comparing mainly dollars, energy, and yields) have found that the more natural methods are definitely economically feasible, especially with higher fuel, chemical, and fertilizer costs (Johnson, W.A., V. Stoltzfus, & P. Craumer, 1977, Energy Conservation in Amish Agriculture, *Science*, Vol. 198, p. 373-378; Carter, L. J., 1980. Organic Farming Becomes "Legitimate," *Science*, Vol. 209, p. 254-256). And such studies ignore the *higher quality* of crops, plus the incalculable value of improved soil.

Learn. After you have decided that natural systems farming is the way to go—or at least worth trying—what next? Next you should learn as much about the subject as possible, so that you not only will know what you are doing, but why. It is hoped that this book will be a good start in that direction. Other helpful publications can be found in the reading list at the end of this book.

Seek advice. Unless you are very knowledgeable in all aspects of soil chemistry, plant physiology, and animal nutrition and health, and can keep up with the latest developments and research, you should seek the advice of those who are. But you have to be careful. Can you trust the advice they give? Some advice-givers are more interested in selling their product than in really helping the farmer. Others are simply misinformed or caught up in the high technology/low quality "science-will-save-us" syndrome.

Listen carefully to what they say and compare it with what you know about natural systems farming. Do they recommend toxic chemicals to "solve" weed and pest problems, or do they try to get at the real *cause* of the problem? Do they recommend fertilizers that harm the soil and soil life (see Chapter 12)? Do they try to build up the natural systems of the soil-plant-animal-soil cycle? Are they concerned about quality, not just quantity?

Check with their customers and see what results they have gotten. Are they honest in their promises? If they guarantee a certain yield figure, watch out; because no one can guarantee a yield without being able to control the weather.

Costs. Don't be afraid of a reasonable fee for consulting services. After all, you are paying for their "expertise" and advice, based on considerable experience and knowledge. A fee of say $8 or $10 per acre may sound like a lot, but if it gives you just a few more bushels per acre, it pays for itself. And that doesn't include the value of improved soil, higher quality crops, and healthier animals.

It's entirely possible to go on a natural systems program and spend no more than you had been spending for fertilizers and chemicals; and less if you figure in veterinary and grain drying bills.

A step at a time. The steps which should be followed to turn your soil around and begin restoring natural systems are simple. First, your soil should be tested to see where it is now, what its levels of nutrients and other characteristics are. This is like when you have a fever, you put a thermometer in your mouth to see how bad it is. Only when you know what's wrong can you start to identify problems, find causes, and find solutions.

Next, you (or your consultant) should begin correcting the problems by attacking them from two directions: (1) *stop doing wrong things,* and (2) *start doing right things.*

Of course, it's not really that simple, because you have to know what *are* the right and wrong things, and how and when to use the right ones. This is where the advice of a good consultant is invaluable. As we shall see in Chapters 11 and 12, various fertilizers and soil conditioners can have widely different effects on soil and crops, and even the good ones can be mis-used. And then they can interact—one affecting the other—if used together.

Also, don't forget that your soil's problems may have taken decades to develop, so you can't expect to solve them overnight. Depending on your soil type, the weather, and your budget, it may take several years to make real progress. But as we have said, positive results will occur the first year. And every year after that will show improvement.

So that's what you need to do to start changing course. The only thing holding you back is you—so let's *get up and go!*

Chapter 10

Testing, 1–2–3. . .

Does it pay to have your soil tested? How often should soil be tested? Which lab or company is best? Are tests of all the trace elements and the CEC better? Are *computer* test results better than others? Which testing methods give the most accurate results—or the best fertilizer recommendations? And what about plant tissue analysis—does it tell anything? How can you interpret test results? What do they *really* mean?

For those who are not soil chemists, soil and tissue testing can be a pretty bewildering "hocus pocus" operation, so most farmers "leave it to the experts" and follow their advice religiously. But shouldn't those who make their living from the soil know more about what's going on and why, to be able to make more intelligent decisions? The main thing which forms the basis for millions of dollars worth of fertilizer application is soil testing. Let's learn more about it.

We tend to get the impression, from the neat figures on the computer print-out, accurate to one or two decimal places, that soil testing and fertilizer recommendation is an exact science. After all, doesn't the soil testing lab use multi-thousand dollar lab equipment—probably a *flame spectrophotometer* (whatever that is)—plus a COMPUTER?! It's pretty impressive, isn't it?

If you request it, the soil testing lab will probably send you a detailed technical explanation of their testing methods, listing the exact procedure followed, and test reagents and instruments used. That's also impressive.

But did you know that there are different soil testing *philosophies*, or *theories*, very different *methods*, different ways of *reporting the results*, and different *fertilizer recommendations* that result? In a recent (January 1982) AP syndicated news release written by Don Kendall, we learn that the Rodale Research Center sent identical soil samples to 69 private and university soil testing laboratories. The results were astounding. For example, the fertilizer nitrogen recommendations ranged from zero to 230 pounds per acre, and phosphorus recommendations from zero to 150 pounds per acre! In terms of fertilizer dollars, those recommendations mean a difference of $34.50 per acre for nitrogen and $40.50 for phosphorus. In their magazine, *New Farm*, Rodale Press has charged that soil testing laboratories "are largely unreliable in their recommendations," and "the unreliability of the labs is costing farmers money—conservative estimates put the figures at hundreds of millions."

That's pretty serious! Just by following unreliable fertilizer recommendations, farmers may be wasting millions of dollars! Let's find out what's really going on.

Objectives. Why are soil tests run, anyway? Basically, to determine the fertility of the soil, either to know how much fertilizer needs to be added or to predict the yield of a crop. Few farmers would deny that soil testing is necessary, or at least valuable.

Other methods. Actually, there are other ways of evaluating soil fertility besides chemical soil testing. For example, we can actually test-grow plants, either in a small-scale greenhouse test or in full-scale field test plots. A few farmers and many researchers do this. It takes extra time and trouble. Or we can examine growing crops for mineral deficiency symptoms, such as yellowing midvein for nitrogen deficiency and purple leaves for phosphorus deficiency. Trouble is, by the time those symptoms appear, it's about too late to do anything that year. Or, there are lab tests using certain bacteria and fungi to determine certain deficiencies in your soil. These tests aren't used much, however.

Then there are leaf and tissue tests, which determine the amounts of nutrient elements in a certain part of the plant at that time. These tests are increasing in popularity, with rapid field test kits available, besides the more detailed (and expensive) lab tests. But tissue tests, again, may be run too late to correct a deficiency that year. Also, they may not indicate a real soil deficiency. Sometimes a plant will concentrate as much phosphorus in its tissues as possible (giving a high reading on the test) just before the soil is depleted of available phosphorus (S.H. Wittner & F. G. Teubner, *Annual Review of Plant Physiology*, Vol. 10, 1959, p. 17). And if a plant is absorbing but not metabolizing (using) a nutrient it will still show up in a tissue test, even though the plant has no use for it. Test interpretations seldom take nutrient balance and interactions into account.

But most farmers don't take the time or expense to have those tests run. By far the most commonly run tests are chemical analysis soil tests. Again, rapid field test kits are available, but most people send their soil samples in to a soil testing laboratory. Several tests are usually run on all samples, such as pH, nitrogen, phosphorus, potassium, calcium, magnesium, and total exchange capacity or CEC (cation exchange capacity). Other tests can be run as an option, such as percent organic matter, percent base saturation of several elements, and micronutrient (trace element) analysis.

Soil test methods. There is a multitude of different methods used by labs to test for the various elements and properties of soil. For pH, there is no problem; most labs simply mix a soil sample with a certain amount of distilled water and measure the pH with a standard pH meter. Or in an alternate method, a chemical buffer solution is slowly added to obtain a reading.

But for the other properties, exchange capacity and nutrient amounts, there are widely differing methods and philosophies governing soil testing which the farmer never hears about. The layman just assumes that because the test results from one lab show that his soil has 150 pounds of nitrogen per acre or that his soil needs 1500 pounds of lime per acre, that the tests done by another lab would turn out the same. Not so.

Most labs first dry the soil in order to standardize samples. Others use soil with its natural moisture, with the philosophy in mind that they should test under the conditions that the plants experience. Most labs then grind the soil to a uniform fineness. Some do not, realizing that plants do not have grinders. Then the soil sample is extracted; that is, it is soaked for a certain length of time in one or more extracting fluids which dissolve a certain amount of the nutrients present in the sample. Here there is wide variation. We can use water, salt solutions, alkalies, weak acids, and strong acids. Some extracting fluids work better for some nutrients or for certain types of soil. The stronger ones (or a longer extraction time) will naturally dissolve out more of the nutrients. But plant roots do not have strong acids to obtain every last microgram of a nutrient from the soil. Tests using strong extracting fluids are essentially "geology tests"—finding out the total mineral content of the soil.

After the nutrients have been extracted from the soil sample, the extracting fluid is then chemically analyzed to determine the amounts of the various nutrients it contains. This can be done by adding other chemical solutions (reagents) to it and noting certain color changes, or an expensive lab instrument can be used, either an x-ray emission spectrograph or more likely a flame spectrophotomer will be used to give "more accurate results."

The results. So, by whatever method is used, the laboratory comes up with some numerical figures of the amounts of nutrients measured. How accurate they are depends on the lab apparatus and the skill of the lab technician. But how they are reported and the units used can vary widely. Some labs may give pounds per acre (for either the upper 6 inches or 12 inches), others give parts per million, and others milligram equivalents per 100 grams of soil. And then some use elemental terms—phosphorus, potassium, nitrogen—while others use ionic or oxide forms—P_2O_5, PO_4, K_2O, NO_3, NH_4. And what exactly are they measuring? Usually it is called *available* nutrients, but what does that mean? Is it water soluble, exchangeable, extractable, or what? Available to whom—the chemist or the plant? No wonder the farmer gets confused!

Interpretation and recommendations. We have found out how much nutrients a particular sample of soil contains. So what? What does it really mean? How does it relate to growing crops and applying fertilizer? In order for soil analysis results to do the farmer any good, they must be *interpreted.* And here again there are many pitfalls and much variation in philosophies and practices of different labs.

The overall basic philosophy of growing crops is that the plants have certain mineral needs or requirements and that they take a certain amount of minerals out of the soil during a growing season. If the soil is deficient in one or more nutrients, they will be limiting factors in plant growth. Any available nutrients removed by plants must be replaced if they are in short supply. So far, so good. But there are additional things to consider. First, soils and plant growth needs change *during* the growing season. A plant's need for most nutrients is greatest during the middle and later part of its life, when most growth occurs, plus seed or fruit development. Also, nutrients can become available *during* the growing season and over the winter due to chemical and biological activity in the soil. Then too, the weather conditions during the year affect these variables considerably. Virtually all soil testing operates on the simplistic theory of merely replacing the nutrients removed by the crop or fertilizing to obtain a certain yield. These ideas are just not up-to-date.

Then too, there are different goals and philosophies that the farmer can follow. Much of the time, a farmer applies fertilizer to achieve a certain yield goal, such as 150 bushel corn, or 50 bushel soybeans. Also, one can be on a maintenance fertilizer program, seeking to stay at a certain level of fertility, or one can be on a corrective, or building program, seeking to improve the soil. All of these factors greatly influence the interpretation of test results and fertilizer recommendations.

But there is still one more thing—the amounts of nutrients found in a soil test must be correlated to actual crop response in the fields. In other words, test results have to be calibrated.

Calibration. This is done by conducting plant growth tests, both under greenhouse and field conditions. Different soil types must be tested, different hybrid varieties, and in different climates and latitudes. This extremely important activity takes much of the time and expense of university and industry experimental farms. It is a crucial part of the whole fertilizer-soil test procedure, because this is the major factor upon which fertilizer recommendations are made. Yet we read in the standard textbook, *Soil Fertility and Fertilizers*, 2nd Edition, by S. L. Tisdale and W. L. Nelson (1966), "Much additional work in calibration is needed in the United States as well as in other countries" (p. 481). In 1981, the University of Wisconsin Soil and Plant Analysis Lab announced numerous revisions in its recommendation levels *(Proceedings of the 1981 Fertilizer, Aglime and Pest Management Conference,* p. 91-99, Volume 20).

So we see that one of the main pillars supporting the structure of soil testing and fertilizer recommendations is rather shaky. Again, the actual fertilizer/lime recommendations are based on many possible variables—field tests, soil types, soil test results, maintenance versus corrective goals, *and* the basic approach and philosophy of the lab's director(s) toward soil fertility and soil management.

No Foundation

Could the foundation of the fertilizer recommendations of modern agriculture be non-existent? Yes, believe it or not. Virtually all university, extension, and fertilizer company recommendations are based on the *balance-sheet theory*, first set forth by the German chemist, Justus von Liebig, in 1840. His theory, arrived at after studying the growth of plants and analyzing the soil they grew in, states that plant growth and yields are in "exact proportion" to the amount of mineral nutrients they take out of the soil.

But von Liebig did not take in to account living organisms in the soil; he assumed that soils are like lifeless storage bins of water and inorganic mineral elements (*Soil*, U.S.D.A. Yearbook of Agriculture 1957, p. 3-4).

The balance-sheet theory won the day, partly due to von Liebig's prestige and argumentative skill, and it today so permeates the thinking of the "experts" that they can hardly imagine the overwhelming importance of organic matter and living soil organisms. The experts' figures and recommendations of how much fertilizer each crop "requires" (such as 115 lbs/A nitrogen, 40 lbs/A phosphate, and 116/A potash for 80 bushel corn) follow the balance-sheet theory right down the line. In other words, put in what the crop takes out.

Yet von Liebig's balance-sheet theory has definitely been proven false (*Soils*, U.S.D.A. Yearbook of Agriculture 1957, p. 7); von Liebig himself admitted it was false before his death.

How much simpler and cheaper it is to let nature's volunteer army go to work and supply crops with the kind and balance of nutrients they need?

Exchange Capacity

Plant nutrients in the soil solution (soil water with its dissolved chemicals and gases) are in ionized form, with either a positive or negative charge. Clay and organic colloid particles have a negative charge on their surfaces, and hold (adsorb) positively charged ions (cations, such as NH_4^+, Mg^+). Some ions are held tighter than others, thus they are not as leachable by water.

In the root zone, nutrient cations can exchange places with other cations, such as hydrogen ions (these are what cause soil acidity), which come from plant roots, from the respiration of microorganisms, and from decay of organic matter. (Anion exchange can also occur, but most anions, such as NO_3^-, are directly absorbed from the soil solution.)

Cation exchange capacity (CEC) or base exchange capacity, is a measure of the cation holding ability of clay or humus. It is usually measured by soil testing laboratories. Another related measure is base saturation, which is the percent of CEC saturated by cations; that is, how much of the soil's capacity is being used.

While these measures have their value, they also have their limitations. For example, the soil is constantly changing, and the addition of organic matter can greatly change the CEC. Also, in soil tests, not all cations are measured, so the CEC is only a part of the total picture.

Philosophy. That last factor is a "whole new ball game!" But it is another of the pillars holding up the structure, perhaps the very most important one. As we have hinted earlier, some people believe testing and recommendations should be based on conditions as nearly as possible approaching those found in nature, where the plants actually grow. They believe that there are natural biological systems and laws operating that will produce optimal crop growth and yields with a minimum of human interference. At the opposite extreme are people who believe that modern science and technology will provide the answers, and that growth and yields are vastly improved by the use of synthetic fertilizers, herbicides, and pesticides. Then there are others between these two extremes. A proponent of one soil testing method, double equilibration with soil buffers, has called the "rival" method, cation exchange capacity (CEC) analysis, a fad, and says that its use for making fertilizer and liming recommendations where it is not appropriate has led to absurd recommendations (J. Hanway, *Proceedings of the Third Soil-Plant Analyst Workshop*, 1970, p. 2).

But do you see how one's overall philosophy toward soil management can dictate which and how much fertilizer to recommend? Some commonly used fertilizer materials are actually toxic to seedlings, roots, and beneficial soil organisms, not to mention herbicides and pesticides. Some destroy soil humus.

Computers. The soil test results are "translated" into fertilizer and lime recommendations, generally by a mathematical calculation based on previous field test calibrations, yield goal, maintenance versus corrective goals, and overall soil management philosophy. Most labs use a computer to do the calculations and make the print-out to save time (and impress the customer?). Many labs simplify the interpretation of test results by grouping the numerical readings into several categories or levels of soil nutrients, such as high-medium-low or very high-high-medium-low-very low. Others do this by drawing a graph of nutrient levels. These practices may be helpful, but if a nutrient level is close to a cut-off line, it could be misinterpreted (for example, a level just above the "low" category).

What else? But most labs do not include in their standardized formula for calculating fertilizer recommendations such important factors as the past and future crops grown and planned for the field, and the present organic matter or humus in the soil. For example, most labs recommend adding from 1.0 to 1.25 pounds of "nitrogen" for each bushel of corn expected. They ignore the fact that if soil contains abundant humus and is healthy and alive with beneficial microorganisms, organic matter will break down during the growing season to supply over 50 pounds of nitrogen per acre. Also, if legumes have been grown on healthy soil, nitrogen-fixing root nodules will add considerable nitrogen (about 50-200 pounds per acre). Even lightning and rain add a surprising amount of nitrogen to soil, about 15 pounds per acre. Very few labs test for the two kinds of nitrogen, nitrate and ammoniacal. The growth of many crops is very different under fertilization with one or the other. And don't forget that the lab recommendations are based on standard formulas and field calibration tests. How were those tests run? What fertilizer materials, herbicides, and pesticides were used? Was the soil full of living microorganisms or sterile and dead? Was it aerobic or anaerobic? Will *your* fields behave the same way? These factors play a *tremendous* part in determining the yield and quality of crops.

What then? Does this mean that soil tests are totally unreliable and worthless? Absolutely not. Soil tests are a *tool*, an aid in (1) finding out where your soil is now, and (2) planning for the future. But they have their limitations and they do not consider all the important factors that influence crop growth and yield. They need to be used intelligently and cautiously.

How often? How often should your soil be tested? Most labs or agricultural consultants recommend testing at most once a year—or even 2, 3, 4, or 6 years. And most farmers certainly don't test their soil very often. They're just too busy trying to keep up. But think about this: soil conditions and nutrient levels change considerably during a growing season, even from week to week and day to day. Different crops and fertilizers affect soil conditions. Tillage methods and weather do too. Soil pH can change by at least two points (100 times more acidity) during a growing season. Sometimes nutritive deficiencies can be corrected during that growing season if detected soon enough.

Doesn't it make more sense to have your soil tested more than once a year? Like (1) before planting to plan fertilizer application for the year, (2) during the growing season to detect possible problems, and (3) in the fall to see what nutrients have been removed and to plan for next year. It tests are done only every year or two, how can intelligent fertilizer choices be made? And what time of the year was the sample taken? It makes a great deal of difference. Your crops and money are too valuable to gamble.

Which tests? Is it necessary to have all the available tests run on your soil every time, including trace elements? No, in fact most of the trace element tests are not very reliable for predicting crop response, anyway. It is most important to correct the amounts and balances of the major elements first, namely calcium, phosphorus, potassium, and nitrogen, both nitrate and ammoniacal. After these are pretty much in line, then it will be worth while to check trace elements, unless of course obvious trace element deficiency symptoms are visible in plants or you have a problem type of soil that is known to have trace element deficiencies.

Sampling. Soil test results are only as good as the samples being tested. Since only a very small amount of soil from a large field is actually being analyzed, one should be very careful that truly representative samples are taken. If the soil test lab you have chosen to use has any specific instructions, be sure to follow them. In general, a composite sample is taken from a field; that is, several small samples from different parts of the field are mixed together to give an average or representative sample. In most cases this works well, but if a field has great differences in soil type, contour, or past treatment (crops, fertilizers, lime, manure), separate samples should be taken. Usually, samples should be taken from the plow layer, 6-10 inches in depth, or from the upper 6 inches if a non-plowing tillage system is used. Avoid sampling in a fertilizer band, in poorly drained areas, dead furrows, corners, or any other areas that are not representative. Stay at least 50 feet away from roads, fence rows, and barnyards.

Any tool can be used to dig down. Standard soil probes (tubes) and augers work well, but a shovel will do. Do not touch the samples with rusty iron tools or your fingers. Many (10-30) small samples from all over the field (you can use a zig-zag path as you cover the field) should be taken and mixed together to give a composite sample of 1/2 to 1 pound (1 to 2 cups). Place each composite sample in a separate paper or plastic bag with a clear label giving your name, address, and the sample designation (number or letter, such as Field A, or Field 3. Designate different samples within a field, such as A-1, A-2, A-3).

It is very important that you make a map of your farm and show where the samples were taken in each field. For consistent long term testing, samples should be taken from the same area each year. Keep accurate records of the history of each field: crops grown, fertilizers applied (formula and amount), lime, manure, and other additions (herbicides, pesticides). Also, a record of problems and yields would be helpful.

Beware. Should you really trust your soil, crops, and precious money to soil testing-fertilizer recommendation procedures that may be unreliable, or based on questionable philosophies and field tests?

The ideal situation would be for the farmer to be knowledgeable in soil chemistry, plant physiology, fertilizers, and all the methods and procedures used by soil test labs. But such a person is a "rare bird" indeed. The farmer must rely on the expertise of others. He should certainly be careful.

CHAPTER 11

Fertilizers— What's in the Bag?

Unfortunately, fertilizers are still licensed, promoted, and sold under the outmoded "magical" N-P-K concept, but anyway, when you buy a bag of, say 9-23-30, DO YOU KNOW WHAT'S IN THE BAG??

Chances are, you don't. What, chemically speaking, makes up the 9 pounds of "N"? What accounts for the 23 pounds of "P"? And what about the 30 pounds of "K"?

Did you know there are hundreds of different ways to mix a 9-23-30 fertilizer? That there are dozens of *sources* of nitrogen, phosphorus, and potassium? That some types of fertilizers and soil conditioners can actually *harm* your soil and crops or cause crops to be unhealthy? That two different sources of nitrogen can behave very differently in the soil, having opposite effects on pH and on plant growth, one causing vegetative growth and the other promoting seed development? More about that in the next chapter. In this chapter, we want to examine some of the commonly used sources of plant nutrients and see how fertilizer formulas are figured. *You need to know what's in the bag*!

Labels. Federal and state fertilizer laws require that fertilizers be labeled to show the guaranteed nutrient content of at least the "big three"—N, P, and K—and sometimes secondary and trace elements. So your first source of information should be the label on the bag or tank.

The percent of plant nutrients (or pounds per hundred) listed on the label is called the *grade* or *analysis*. At least the three elements, nitrogen, phosphorus, and potassium, are often referred to in a ratio, such as 10-5-10 or 2:1:2, although not all three may be present, as in 0-20-0. Any additional elements can be added, but must be identified, such as 10-5-10-2S, which contains 2% sulfur.

The analysis can be given either for the P and K oxides (called phosphoric acid or P_2O_5 and potash or K_2O) or for the elemental forms, phosphorus and potassium. Most labels will have both, but the tendency is to label on the elemental basis. One can easily be converted to the other: % or pounds of P_2O_5 multiplied by 0.44 gives P (in reverse, % or pounds of P multiplied by 2.29 gives P_2O_5), and % or pounds of K_2O multiplied by 0.83 gives K (while % or pounds of K multiplied by 1.2 gives K_2O).

In the bag? However, what you are getting in the bag or tank is NOT pure elemental nitrogen, phosphorus, or potassium, and not even P_2O_5 or K_2O. If you look carefully, the label should read P_2O_5 *equivalent* or K_2O *equivalent.* The material you buy *does* contain the stated percentage of nitrogen, phosphorus, and potassium, but those chemical elements are in *combined form* as chemical compounds, either as inorganic or organic compounds. They are never in the oxide forms of P_2O_5 or K_2O—those are just standards for comparison and labeling purposes. And as we said earlier, there are dozens of different commonly used sources of plant nutrients, each with its own properties and effects on soil and crops.

Physical form. Also, fertilizers can be prepared in several physical forms, which have their own specialized equipment for storage and application. These are (1) *gas*, mainly anhydrous ammonia, a source of nitrogen; (2) *dry solids* (either powders, crystals, or pellets), which can

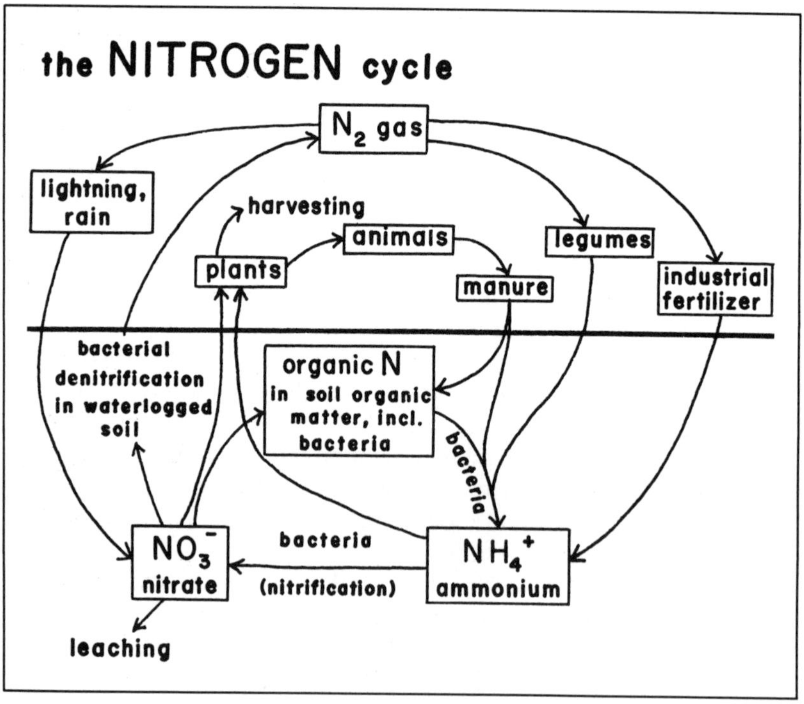

contain one or more nutrients; (3) *liquids*, one or more nutrients in water solution; and (4) *slurries* and *suspensions*, combinations of dry and liquid components. In general, there is no significant difference in the effectiveness of one form of fertilizer over another. The big differences are in the chemical *source*. So let's look at common sources of N, P, and K . . . at a "shopping list" if you will, of possible nutrients to choose from.

Nitrogen. The air around us is mostly elemental nitrogen—78%, but in this form, plants cannot use it. Plants mainly use nitrogen in the nitrate (NO_3^-) and ammonium (NH_4^+) forms. Soil organisms convert ammonia (NH_3) and ammonium forms to the nitrate form, which readily leaches into the groundwater (the conversion processes are called nitrification). Waterlogged soil can also result in nitrogen loss through a more or less opposite series of microbial changes called denitrification. Nitrogen from the air can be trapped, or fixed, by nitrogen-fixing bacteria which can either live independently in the soil or in the root nodules of legumes. That is, provided the soil environment is favorable for beneficial soil organisms, and not sterilized by toxic substances. Additional nitrogen is supplied by lightning and rain (but rain also causes leaching of nitrates).

The greatest need of nitrogen in typical plants is during their time of maximum vegetative growth. Since nitrogen can be so easily lost from the soil, an effective fertilization strategy is to provide a nitrogen source or sources that are non-leachable and available over long periods of time, or to use split applications during the growing season.

A look at the accompanying table will reveal great diversity in nitro-

PRINCIPAL NITROGEN SOURCES (compiled from several sources)

Material	Chemical Formula	% Nitrogen	Form	Acidity (−) or Alkalinity (+)*	Advantages, Disadvantages
ammonia	NH_3	82 (gas)	gas (anhydrous ammonia),	−148 (gas)	Low price, high N analysis; non-leachable after changes to NH_4^+ on soil colloids.
		20-24 (liq.)	liquid (aqua ammonia)	−36 (liq.)	Kills soil organisms; hardens soil; can be lost to air; at first causes alkalinity, then acidity.
urea	$CO(NH_2)_2$	42-46 (dry) 20 (liq).	dry, liquid	−71	High N analysis; non-leachable after changes to NH_4^+ on soil colloids. Can change to ammonia and escape into air if applied on surface.
ammonium nitrate	NH_4NO_3	33-34	dry	−62	High N analysis; supplies two types of N. Nitrate is leachable.
ammonium sulfate	$(NH_4)_2SO_4$	20.5-21	dry	−110	Non-leachable N; also supplies sulfur. Low N analysis; medium price; causes acidity.
sodium nitrate (nitrate of soda)	$NaNO_3$	16	dry	+29	Immediately available N. Low N analysis; leachable N.
potassium nitrate	KNO_3	13.5-13.8	dry, liquid	−23	Immediately available N; also supplies potassium. Low N analysis; leachable N.
calcium nitrate (nitrate of lime)	$Ca(NO_3)_2$	15.5-17	dry, liquid	+20	Immediately available N; also supplies calcium. High price; low N analysis; leachable N.

PRINCIPAL NITROGEN SOURCES (continued)

Material	Chemical Formula	% Nitrogen	Form	Acidity (−) or Alkalinity (+)*	Advantages, Disadvantages
mono-ammonium phosphate (MAP)	$NH_4H_2PO_4$	11-13	dry	−58	Non-leachable N; also supplies available phosphorus. Low N analysis.
diammonium phosphate (DAP)	$(NH_4)_2HPO_4$	18-21	dry	−70	Low price; non-leachable N; also supplies available phosphorus. Can injure seedlings (high ammonia).
ammonium thiosulfate	$(NH_4)_2S_2O_3$	12-18.9	liquid	neutral	Non-leachable N; also supplies sulfur. Can injure plants if used wrongly.
animal manures	various organic compounds	0.6-2.0	dry, liquid	slightly alkaline to slightly acidic	Non-leachable N, available over long period; improves soil structure; also may supply many other elements. Low N analysis (but N content can grow with biological activity); large amounts can compete with crops, or (animal manues) can cause salt damage to roots.
green manures	various organic compounds	0.7-7	plowed under	neutral	
sewage sludge	various organic compounds	2-6.5	dry, liquid	slightly acidic	
tankage	various organic compounds	3-10	dry	slightly alkaline to slightly acidic	
bloodmeal	various organic compounds	13-15	dry	slightly acidic	

* Acidity = pounds of $CaCO_3$ needed to neutralize acid per 100 pounds of material. Alkalinity = equivalent needed to neutralize alkalinity per 100 pounds of material.

Ammonia and Soil Don't Mix

Quotes from an article by M.M. Mortland entitled "Reactions of Ammonia in Soils" in *Advances in Agronomy*, Volume X, 1958:

"The immediate effect of an injection of anhydrous ammonia or application of an aqueous solution of ammonia, is to create a very alkaline reaction in the vicinity of the application."

He then gives several examples:

original pH	*pH after ammonia*
6.0	8.1
5.0	6.2, 7.1
4.9	6.5, 8.0
7.3	8.8, 9.5
8.1	10.1, 10.6

"Ammonium hydroxide has a strong solvent action, as well as a hydrolytic action on organic matter. Thus applications of anhydrous ammonia or aqueous solutions of ammonia which result in localized regions of high concentration in the soil might be expected to result in solubilization and hydrolysis of certain fractions of the soil organic matter.

"Some published information indicates that ammonia may have an effect on soil structure. Smith (1954) has suggested that anhydrous ammonia is a "soil conditioner" in its effects on the physical properties of soil. His contention is that organic compounds are dissolved in aqueous solutions of ammonia, then coat the soil aggregates which become water-stable upon drying. Other results indicate the formation of a poorer soil structure. That NH_4OH has dispersive effects on colloids under certain conditions is well known. . . . It tends to coagulate soils having neutral or alkaline reaction, whereas it tends to disperse or deflocculate acid soils. . . . Such changes in properties of constituent minerals alter the physical properties of the soil.

"Ammonia, however, may be toxic to living organisms. It may exert a toxic or even lethal effect on the cells since at high concentration ammonia is a cell poison. In the application of anhydrous ammonia to soils, the tops of plants are affected wherever the free gas contacts them. Some effect on the roots when crops are side-dressed may also be expected. The extent of injury may be quite variable.

"Since ammonia is toxic to cells at high concentrations, it might be expected to affect the soil microbial population. . . . Eno and Blue (1954) found that anhydrous ammonia applications to Aredondo loamy fine sand and Lakeland fine sand decreased the number of soil fungi. Both bacteria and fungi were decreased the first day. By the tenth day fungi were still reduced in number while bacteria had increased and were 6 to 25 times as numerous as in the check."

gen sources, each with its advantages and disadvantages. There is no one "best" nitrogen source. It all depends on what you need and what your local conditions are—your soil type, soil test results, crops previously grown and planned, weather, tillage methods, available machinery, and budget. Only one of the sources cannot be recommended: anhydrous ammonia. Its disadvantages of killing soil life and hardening soil by destroying organic matter far outweigh its advantages. In fact, the use of soil injection of anhydrous ammonia was perfected during World War II, when it was used to make soil hard to build landing strips for bombers (*Acres U.S.A.*, May 5, 1981).

Phosphorus. Soils have a super-abundance of phosphorus, 1000-2000 pounds per acre. The only trouble is, most of it is unavailable to the plant. From 40-75% is chemically "locked up" in the form of compounds with calcium, iron, aluminum, manganese, and other elements. These inorganic forms of phosphorus are for all practical purposes unavailable to the plant on a short-term basis because the phosphorus is so tightly bound. High or low pH greatly reduces the availability also.

From 25-60% of the soil's phosphorus is in the form of organic compounds and part of the cells of soil organisms. By the action of soil organisms on organic and inorganic forms of phosphorus, most of the plant's supply of phosphorus is slowly made available.

Which would you buy, a bag of 0-20-0 or a bag of 0-2-0? The fertilizer label proudly tells you how much phosphorus is "in the bag" in a *water soluble form*, which is the form plants can use. But what the label doesn't tell you is that within a matter of *a few hours,* nearly all (80-90%) of the water soluble phosphorus has been chemically converted to water *in*soluble forms, and is thus not immediately available to your plants (*Profitable Management of Wisconsin Soils,* 1972, p. 35). You have to read technical fertilizer books to find that out. Yet most of the phosphorus sources used in commercial fertilizers have started out as insoluble rock phosphate and have been chemically treated to produce a large percent of soluble phosphate to make the label analysis look good. All of which expense (out of your pocket) is undone in a few hours!

A plant's need for phosphorus is fairly constant throughout its growth period. Needs are as high during young growth as later, but the root system may not be developed enough to supply the need. Plenty of nitrogen aids in phosphorus uptake. And don't forget about those beneficial soil fungi, the mycorrhizae, which live in and on plant roots and help in phosphorus absorption (see Chapter 3).

The accompanying table of phosphorus sources shows a variety of materials, with varying properties. Again, the extra expense to gain increased phosphorus solubility in the first six materials is largely wasted.

PRINCIPAL PHOSPHORUS SOURCES (compiled from several sources)

Material	Chemical Formula	% Phosphorus	% P_2O_5	Form	Acidity (−) or Alkalinity (+)*	Advantages, Disadvantages
superphosphate (monocalcium phosphate)	$Ca(H_2PO_4)_2$ + $CaSO_4$	7-9.6	16-22	dry	neutral	Also supplies calcium & sulfur (also contains calcium sulfate). High Price.
triple superphosphate (monocalcium phosphate)	$Ca(H_2PO_4)_2$	19.6-20.5	45-47	dry	neutral	Also supplies calcium. High price.
mono-ammonium phosphate (MAP)	$NH_4H_2PO_4$	8.7-21	20-48	dry	−58	Also supplies non-leachable nitrogen.
diammonium phosphate (DAP)	$(NH_4)_2HPO_4$	20-23.1	46-53	dry	−70	Also supplies non-leachable nitrogen. Can injure seedlings (high ammonia).
ammonium polyphosphate	$NH_4H_2PO_4$ + $(NH_4)_3HP_2O_7$	15-26.6	34-61	liquid, suspension	slightly acidic	High P analysis; makes some trace elements more available (Fe, Zn, Mn); also supplies nonleachable nitrogen. More slowly available.
phosphoric acid	H_3PO_4	23-24	52-54.5	liquid	−110	High P analysis. High price.
hard rock phosphate (calcium fluorapatite & hydroxyapatite)	$3\ Ca_3(PO)_2 \cdot CaF_2$ + $3\ Ca_3(PO_4)_2 \cdot Ca(OH)_2$	11.5-17.5	26-40	dry	slightly alkaline	Also supplies trace elements (Fe, Mg, B) and calcium. Low availability (but increases with biological activity).
soft rock phosphate (colloidal phosphate, rotten phosphate)	same as above + clay	8-11	18-25	dry	slightly alkaline	Higher availability than hard rock phosphate; also supplies trace elements (Fe, Mg, B) and calcium.
basic slag	contains $CaSiO_3$ & P_2O_5	3.5-8	8-18	dry	alkaline	Also supplies trace elements (Fe, Mg, B, Mo, Cu, Zn, Mn) and calcium.
bone meal	various organic compounds	9.7-13	22-30	dry	slightly alkaline	Also supplies nitrogen.

* Acidity = pounds of $CaCO_3$ needed to neutralize acid per 100 pounds of material. Alkalinity = equivalent needed to neutralize alkalinity per 100 pounds of material.

Phosphorus

Although plants and animals need many different mineral elements, some are needed in greater quantities than others, and phosphorus is one of those major elements, one that is often seriously deficient—in forms plants use—in today's soils. Phosphorus is essential for all life, both plant and animal. It is part of the molecules that make up genes (DNA) and those that carry genetic information from the genes to the "protein factories" of the cell (RNA). Also, phosphorus is a necessary component of several kinds of molecules that carry and temporarily store cellular energy (high energy phosphates), both during photosynthesis, when the sun's energy is trapped, and later during respiration, when food energy is released.

When a plant is deficient in phosphorus, it will continue to carry on photosynthesis for a time, but its energy production and food transfer activities will "shut down." Therefore, sugars from photosynthesis will accumulate, but will not be changed into proteins and fats. Sometimes a purple pigment (anthocyanin) will accumulate in the leaves. The reserve supply of phosphate ions in the cells will quickly be used up to form the organic phosphate carrier and genetic molecules, and once that happens, phosphorus is no longer available to new growing cells where it is vitally needed (*Phosphorus for Agriculture: A Situation Analysis*, p. 10).

There are several factors that influence soil phosphorus availability to plants. Since most soils have plenty of phosphorus in chemically tied-up forms, we should pay special attention to factors that can overcome this problem. In soils of low organic matter content, lower pH (below 6.5) causes much phosphorus to become tied up in iron and aluminum compounds. However, high organic matter counteracts this and gives good P availability down to pH 5 or 6. Several studies have proved that organic matter and compost increase the availability of not only the phosphate already in the soil, but also of added rock phosphate fertilizers. Organic acids produced by microorganisms are at least partly responsible. In one experiment, application of 80 pounds per acre of soluble phosphate alone tripled the uptake of phosphorus by crops, but mixing the same amount of phosphate with organic matter (spent brewer's hops) increased uptake *five* times (*Soil Science*, 1952, vol. 73, p. 173; *Proceedings of the Soil Science Society of America*, 1957, vol. 21, p. 278). The role of root mycorrhizae in increasing phosphorus absorption has already been covered in Chapter 3.

Optimum soil moisture, temperature, and aeration are very important for phosphorus availability because they increase microbial activity in the soil, plus roots need optimum conditions for effective nutrient uptake. Just an increase in soil temperature from 41° to 80°F increased the P uptake by corn almost four times and the yield (dry matter) over 5 times (*Proceedings of the Soil Science Society of America*, 1961, vol. 25, p. 369).

Besides promoting general growth and yield, phosphorus also directly or indirectly increases crop quality, drought resistance, disease resistance, nitrogen fixation, and decreases maturity time, especially in cool temperatures (*Phosphorus for Agriculture: A Situation Analysis*, 1978, p. 42-111).

Potassium. All soils except sands have enormous amounts of potassium, but like phosphorus, most (98-99%) is unavailable, being "locked up" in soil minerals. What potassium that is available is held on clay or humus colloids, which is why sandy soils cannot hold much and must have an adequate amount added each year. Except for sandy soils, the mineral reserve should supply most potassium needs. If fertilizer potassium is added, any excess not held by soil colloids can leach away.

A plant's need for potassium starts out small and grows continually. In corn, about one-fourth of the potassium goes to the grain and three-fourths are in the rest of the plant.

Only several fertilizer materials are commonly used (see accompanying table), and one of those, potassium chloride (muriate of potash, 0-0-60 or 0-0-62), which is used far more than any other, cannot be recommended because it contains 47.6% chlorine. Some crops are sensitive to chloride, especially potatoes and tobacco—also corn, soybeans, sweet potatoes, strawberries, cucumbers, peppers, citrus, and blueberries—in amounts over 100 parts per million of chloride. Muriate of potash is a high salt fertilizer (see Chapter 12, p. 135-136). High salt concentrations injure or kill roots *and* interfere with the normal life activities and reproduction of beneficial soil microorganisms. Some agronomists, extension agents, and fertilizer dealers recommend levels of muriate of potash for alfalfa of 200 to 800 pounds per acre. Assuming that the plow layer of soil weighs about 2 million pounds, that's about 50-200 parts per million chloride! Is it any wonder that most alfalfa roots no longer have nitrogen-fixing bacterial nodules? The "experts" and the textbooks they write say that chloride forms salts with calcium, magnesium, and sodium, and quickly leaches out of the soil. One study showed that it takes two years for the chloride to leach below two feet (*Better Crops with Plant Food,* Vol. 66, Summer 1982, p. 26-28). Is this "quickly"? According to the book, *Fertilizer Technology & Use,* 2nd ed., 1971, p. 331, the use of muriate of potash in the row along with the seed may cause salt injury to the seedling; on clay soil about 18 pounds per acre may be injurious, but the danger level is lower on sandy soils.

Finally, according to C. Walters, Jr. and C. J. Fenzau in *An Acres U.S.A. Primer* (1979, p. 192), excessive chlorine will cause a potassium deficiency in the soil. How then can a fertilizer be recommended that is made up of BOTH potassium and chlorine?! It sure makes for a lot of repeat business!

Calcium and pH. Why do farmers lime their soil? A loud chorus answers, "To correct acidity," or "To sweeten the soil." Almost everyone thinks of the purpose of liming materials as being to adjust soil pH, not to

PRINCIPAL POTASSIUM SOURCES (compiled from several sources)

Material	Chemical Formula	% Potassium	% K_2O	Form	Acidity (−) or Alkalinity (+)*	Advantages, Disadvantages
potassium chloride (muriate of potash, kalium potash)	KC1	49.8-51.5	60-62	dry, suspension	neutral	Low price. Chlorine kills soil organisms, can harm plants (contains 47.6% chlorine).
potassium sulfate (sulfate of potash)	K_2SO_4	39.8-44	48-53	dry	neutral	Nearly chlorine-free (2%); also supplies sulfur.
potassium nitrate (nitrate of potash)	KNO_3	37-38.7	44-46.6	dry, liquid	−23	Nearly chlorine-free (1%); also supplies nitrogen. High price.
sulfate of potash-magnesia (potassium magnesium sulfate, sul-po-mag)	$K_2SO_4 \cdot MgSO_4$ + $MgCl_2 + H_2O$	18.3-19.2	22-23	dry	neutral	Nearly chlorine-free (1%); also supplies magnesium & sulfur.
granite dust; greensand, glauconite	variable: potassium aluminum and iron silicates	3-5 (granite) 6-7 (greensand)	3.6-6 7.2-8.4	dry	slightly alkaline	Potassium released slowly; also supplies trace elements (Fe, Mg).
animal manures	various organic compounds	0.3-0.8	0.36-1.0	dry, liquid	slightly alkaline to slightly acidic	Potassium released slowly; also supplies non-leachable nitrogen. Large amounts can compete with crops or cause salt damage to roots.
sawdust	various organic compounds	6-7	7.2-8.4	dry	slightly acidic	Potassium released slowly. Large amounts can compete with crops.
wood ashes	various organic compounds	7-18	8.4-21.6	dry	alkaline	Also supplies trace elements. Can injure seedlings.

* Acidity = pounds of $CaCO_3$ needed to neutralize acid per 100 pounds of material. Alkalinity = equivalent needed to neutralize alkalinity per 100 pounds of material.

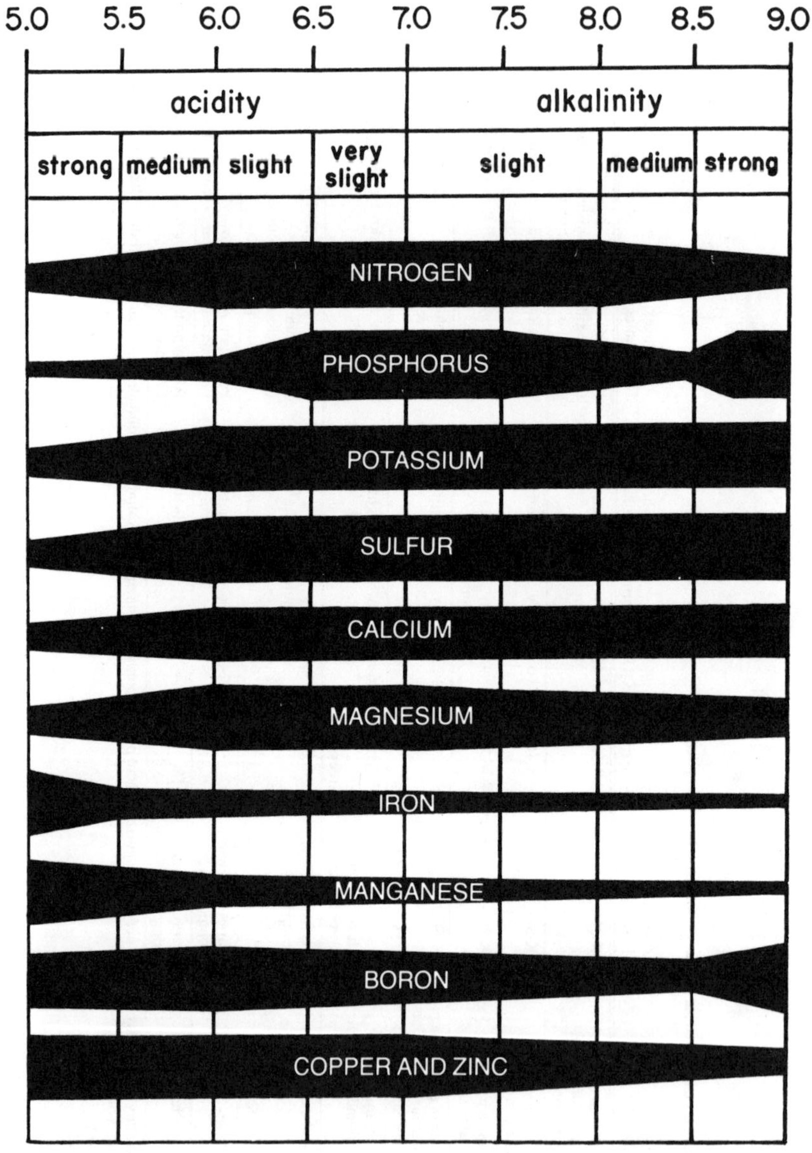

Availability of Elements to Plants at Different pH Levels for Typical Soils

fertilize. In fact, materials used in pH and salinity control are called soil amendments, not fertilizers.

To measure acidity and alkalinity, we use the pH scale, a numerical scale running from 0 (most acid) to 14 (most alkaline), with 7 as neutral being in the middle. A change along the scale of one whole number (from 6 to 5, or from 7.5 to 8.5) means a change in acidity or alkalinity of 10 times; thus a pH of 4 would be 100 times more acid than a pH of 6.

Soil pH is usually from 4 to 10, but most crops do best under slightly acid conditions (6.0-6.8), although most crops can do well under a fairly wide range of pH, provided they receive an adequate balance of nutrients. The pH does affect the availability of nutrient elements, and at very acid or alkaline conditions, nutrient deficiencies and toxicities can occur (see accompanying graph). Aluminum and manganese are more soluble at very low pH and can be toxic. Soil pH also plays an important role in determining the ability of beneficial soil organisms to survive, for while fungi can tolerate low pH (below 6), most bacteria cannot. Thus, ordinary humus-forming bacteria, as well as nitrogen-fixing bacteria can be hampered.

So, "correcting" soil acidity can be necessary and important, if the soil is very acid, but there are other things to consider. Most "experts" do not consider that the pH can change greatly during the year. In an actual example of soil from a farm in southeastern Minnesota, the pH was 6.4 on May 28, 1982, and 4.8 on July 19, 1982.

Soil pH can change because of rains, because of microbial and root activity, and as we see in the tables in this chapter, because of using one or another fertilizer. For example, anhydrous ammonia causes a strong alkaline pH at first, then later becoming acid, while superphosphate causes an acid pH. These pH changes can be detrimental to crops and soil life.

The accompanying table of liming materials and calcium soures reveals some that are more desirable than others. Most often, ground limestone from local quarries is used, at a very low cost. But there are two kinds of limestone, calcitic and dolomitic, and calcitic is preferred unless your soil is deficient in magnesium. Excessive magnesium (contained in dolomite) can be detrimental by (1) causing soil to crust, reducing aeration, (2) releasing soil nitorgen by causing formation of gaseous nitrogen oxides, (3) causing both phosphorus and potassium deficiencies in the soil, (4) causing effects similar to magnesium deficiency, (5) combining with aluminum to form a substance in plants toxic to livestock, (6) along with low calcium, allowing organic matter to form alcohol and formaldehyde when it decays, killing soil bacteria, (7) interfering with plants' absorption of calcium and potassium, and (8) by taking the place

LIMING MATERIALS AND CALCIUM SOURCES **(compiled from several sources)**

Material	Chemical Formula	% Calcium	Form	Effect on pH	Advantages, Disadvantages
calcitic limestone (calcite, calcium carbonate, high calcium lime)	$CaCO_3$	36-38	dry, liquid suspension	compare with others, below	Low price, rapidly available, high calcium & low magnesium
dolomitic limestone (dolomite, magnesium limestone, aglime)	$CaCO_3 \cdot MgCO_3$	22	dry, liquid suspension	slightly greater neutralizing ability than calcite	Low price. Slowly available; with 12-14% magnesium, which can harm soil & crops.
marl	$CaCO_3$ + clay impurities	variable	dry	10-30% less neutralizing ability than calcite	Low price, little magnesium. Clay impurities.
hydrated lime (slaked lime, calcium hydroxide)	$Ca(OH)_2$	54	dry	1/3 greater neutralizing ability than calcite	Greater neutralizing effect, rapid action, readily available, little magnesium. Easily leached, high price, can cause release of nitrogen & rapid pH change.
lime (unslaked lime, quick lime, calcium oxide)	CaO	71	dry	80% greater neutralizing ability than calcite	Very great neutralizing effect, very rapid action, little magnesium. Can "burn" crops & kill soil life, high price.
calcium sulfate (gypsum)	$CaSO_4 \cdot 2H_2O$	23	dry	no effect on pH	Also supplies sulfur, useful on alkaline soils by getting rid of sodium, little magnesium
calcium nitrate (nitrate of lime)	$Ca(NO_3)_2$	19.5-21	dry	1/4 the neutralizing ability of calcite	Also supplies nitrogen, readily available, little magnesium. High price.
slags, basic slag	contain $CaSiO_3$ ± P_2O_5	about 32	dry	10-40% less neutralizing ability than calcite	Low price, also may supply phosphorus, little magnesium.
superphosphate (monocalcium phosphate)	$Ca(H_2PO_4)_2$ + $CaSO_4$	18-21	dry	lowers pH	Also supplies phosphorus and sulfur.
triple superphosphate (monocalcium phosphate)	$Ca(H_2PO_4)_2$	12-14	dry,slurry	neutral	Also supplies phosphorus.
rock phosphates (hard and soft)	$3Ca_3(PO)_2 \cdot CaF_2$ $3Ca_3(PO_4)_2 \cdot Ca(OH)_2$	33	dry	slightly alkaline	Also supplies phosphorus.

of calcium in plant cells, giving rise to poor quality crops. Also, dolomitic limestone breaks down in the soil very slowly. It may take as long as 18 months before any of its calcium enters plants. If calcitic limestone is not available locally, it will still be worth the extra freight expense to use it if your soil has plenty of magnesium.

The effectiveness of ground limestone depends greatly on its particle size. The finer it is ground, the more rapidly it acts, and the less that is required per acre. The use of coarser particles has the advantage of giving a long-term activity.

By far the greatest benefit of using lime is for its fertilizer value in supplying calcium, not its pH control, since other materials will regulate pH, and in some cases the use of lime will even *lower* pH, not raise it (this can happen when positive aluminum ions react with soil water to release additional hydrogen ions, which cause more acidity). As a specific example, on a farm in northwestern Wisconsin, pH was 6.0 in October 1980; after applying 1000 pounds per acre of calcite lime, the pH was 4.5 in June 1981. A plant's need for calcium has been underestimated by many people. Many experts and books say that calcium deficiencies in soils are rare or non-existent. They base that statement on soil tests which detect

Calcium and Disease

• From article by R. R. Muse & H. B. Couch, in *Phytopathology*, vol. 65, 1965, p. 507-510:

Additional calcium decreased the disease, *Corticium* red thread, in turf grass (creeping red fescue), while high magnesium increased it. ". . . calcium-deficient regimes produced plants with significantly heightened disease susceptibility regardless of the imbalance of other nutrient elements, soil temperature, air temperature, time of plant growth, or host variety."

• From article by S. Shannon, J. J. Natti, & J. D. Atkin, in *Proceedings of the American Society for Horticultural Science*, vol. 90, 1967, p. 180-189:

The calcium content of the seed of snapbeans is inversely related to the incidence of hypocotyl necrosis, a disorder of seedlings in which tissue of the developing root breaks down. High levels of potassium increase severity of blossom-end rot of tomatoes, blackheart of celery, tipburn of cabbage, and cavity spot of carrots.

• From University of Wisconsin-Madison extension paper 81-PF, "Calcium and Magnesium in Potato Production in Wisconsin," by P. E. Fixen and J. L. Walworth, p. 5:

". . . as tuber Ca decreases, susceptibility of bacterial soft rot in storage appears to increase."

all the calcium in the mineral component of the soil. But only a small part of that is readily available to a plant at one time (it slowly becomes available over the years), and many crops are deficient in calcium. Dr. William Albrecht championed the use of lime for its calcium content and showed in his writings that calcium is essential for protein production in plants, it makes phosphorus more available to plants, and it is vital for animal and human health in the skeleton and for milk production. He pointed out that World War II draftees from states with naturally high calcium soils had healthier teeth and lower rejection rates than those from states with low calcium soils. Calcium stimulates growth of soil microorganisms, including nitrogen-fixing bacteria. It is needed in the membranes of all living cells, and is required for proper nerve and muscle activity and blood clotting. The importance of calcium in the soil, in plants, and in animals and humans is very great and a vital part of any soil fertility program. Sill another benefit of using lime is that the carbonate ion (CO_3^-) can eventually break down and release carbon dioxide (CO_2), which is necessary in photosynthesis.

But what's in the bag? Now that we have looked at the most common sources of major nutrients, let's see how a blended fertilizer is put together and how those " magic numbers" on the label are determined.

Again what are you actually getting in the bag or tank? Some people may think they are buying pure plant nutrients, N-P-K (nitrogen, phosphorus, potassium), but that is not the case. You are not even buying P_2O_5 or K_2O, for the label should state that there is a guaranteed amount of P_2O_5 *equivalent* and K_2O *equivalent*. You *are* buying some N, P, and K, but they are chemically combined with other elements. The tables of nutrient sources found earlier in this chapter give the chemical formulas of the non-organic materials. These formulas show what other elements are combined with the nitrogen, phosphorus, and potassium (refer to the Appendix for the chemical abbreviations used). The trend today is away from the confusing use of P_2O_5 and K_2O, toward giving the nutrient content in terms of the elements, N, P, and K. (other elements can also be given after those three, as long as they are identified).

Examples. Let's start with a simple case, a single fertilizer material. One example is potassium chloride (muriate of potash), which is labeled 0-0-60 or 0-0-62 using the K_2O basis. Its chemical formula is KC1, which means that it contains not only potassium, but also chlorine. Its percent of K_2O is 60-62 (49.8-51.5 elemental K) and it is labeled 0-0-60 or 0-0-62, since the chlorine is not normally labeled, although it can have detrimental biological effects in the soil.

Another example would be diammonium phosphate, which is $(NH_4)_2HPO_4$. It contains two major plant nutrients, nitrogen (18%) and phosphorus (20%). Thus its label would be 18-20-0, or if converted to P_2O_5 equivalent, 18-46-0.

Now let's look at a very commonly sold *blended* or mixed fertilizer, 9-23-30. Actually, the formulation can be mixed in many ways by using various fertilizer materials, but is usually made by mixing equal amounts of diammonium phosphate (18-46-0) and muriate of potash (0-0-60). Since these two materials are blended half-and-half, there would be half the numerical amounts per 100 pounds, so:

50 pounds diammonium phosphate =	9 - 23 - 0
50 pounds muriate of potash =	0 - 0 - 30
total	9 - 23 - 30

Next, a more complicated one. Here are two different commercially used formulas for a 6-24-24 fertilizer:

10.0 pounds......	4-44-0	10.0 pounds......	4-44-0
31.25 pounds.....	18-46-0	21.25 pounds.....	18-46-0
11.5 pounds......	0-46-0	8.5 pounds.......	21-0-0
38.75 pounds.....	0-0-62	21.5 pounds......	0-46-0
8.5 pounds.......	0-0-0	38.75 pounds.....	0-0-62

The 8.5 pounds of 0-0-0 in the first formula is a filler used to make a full measure, usually dolomitic limestone.

So what you get in the bag is not all plant food. It is a mixture of chemical elements, in the form of one or more compounds. The label guarantees a certain amount of a few things, mainly the "big three," N-P-K, but what the label *doesn't* say is often more important than what it does say. What else is in the bag? Is it toxic to crops or soil life? Does it harm the soil structure or humus? Is it available when the plant needs it? What actual fertilizer materials were used in blending the formula (if it is a blended fertilizer)? You need to know, because many, if not most commercially sold fertilizers contain harmful materials. The most often used one is muriate of potash, which is 47.6% chlorine. The fertilizer dealer should tell you (if you ask) whether his blends contain chlorine or dolomite. If he doesn't want to say, beware. State fertilizer inspectors are usually more than willing to come to your farm and take a sample of any fertilizer for analysis before it is delivered. If your fertilizer dealer says he is supplying a non-chlorinated fertilizer, this is how you can be sure.

Your best buy. Is the fertilizer with the highest number on the label really the best buy? The best buy in fertilizers should be the one that gives your crops the best nutrition—not just the most N, P, or K at the

lowest cost per acre, but the *right amount* of nutrients, at the *right time* and in the *right forms*—and with *no harmful side effects*. You need to look at the long-term results—the rest of the story—not just the immediate economics.

Let's compare two sources of phosphorus—a vitally important plant nutrient. Superphosphate (0-20-0) contains about 8.7% phosphorus and sells for about $200 per ton (1982 price). The label guarantees that the phosphate is in water soluble form *in the bag*. But the label doesn't say that nearly all the phosphate is converted to insoluble form *in the soil* within hours or a few days. On the other hand, rock phosphate has very little soluble phosphate, perhaps only 2%, so a fertilizer label that reads 0-2-0 doesn't look very impressive, and many farmers wouldn't buy it. It costs only about half as much as superphosphate. But it contains more elemental phosphorus, from 8-17%. Superphosphate is made from rock phosphate. It is treated with sulfuric acid to make the phosphate more soluble (resulting in a larger number on the label and a higher price—and profit to the dealer), but in a few hours, it is nearly all converted back into insoluble rock phosphate form, and your extra money was wasted.

Phosphorus becomes available to plants from the soil's large reservoir of phosphate, slowly as soil bacteria and plant roots produce various acids. Mycorrhiza fungi help feed phosphorus to roots, and animal manures make phosphate more readily available to plants. Tests done at the Virginia Agricultural Experiment Station showed that one ton of rock phosphate applied *once every six years* gives yields equal to superphosphate applied *every year* at 100-280 pounds per acre. The fertilizer costs alone per acre of the two choices (1 ton every 6 years vs. 200 pounds every year) would be nearly equal (about $17 vs. $20), but the extra fuel and labor costs of yearly spreading would tip the balance in favor of rock phosphate. The importance of a healthy, "living" soil is very great in helping to break down the soil's rock phosphate into available form.

In the same way as we have done for phosphorus, we could show that the best buy in nitrogen is not the form with the highest label rating (ammonia with 82% nitrogen) and the best buy in potassium is not the cheapest (muriate of potash), because these fertilizers harm the soil or soil life.

So let's look at the various commonly used fertilizers and see what they do, and whether and how they should be used.

CHAPTER 12

Help or Hindrance?
What Fertilizers Do

The average person—and most farmers—thinks that anything you put on your soil which is called a *fertilizer* must be good and must make your soil more fertile. Must produce better crops. That couldn't be farther from the truth! As we have mentioned in previous chapters, some commonly used fertilizers and soil amendments can actually HARM the soil, can DESTROY soil structure and humus, and KILL beneficial soil organisms. You can't afford to let that happen! You need to know what you are putting on your soil and what it does.

It's elementary. In order to live and carry on photosynthesis, plants need 17 chemical elements. Three of them—carbon, hydrogen, and oxygen—are obtained from water (H_2O) and the carbon dioxide (CO_2) in the air. Except for the nitrogen obtained from the air by the nitrogen-fixing bacteria in legume root nodules, plants obtain most of their needed nutrients from the soil (there is some evidence that mineral nutrients can be taken from dust particles in the air; also trace amounts of ammonia and sulfur dioxide gas in the air can be used by some plants).

The three elements, carbon, hydrogen, and oxygen, partly are used as the main building blocks of most of the carbohydrate, fat, and protein molecules that make up the solid parts of cells. The basic formula for photosynthesis is a combination of carbon dioxide and water to form sugar: $CO_2 + H_2O \rightarrow C_6H_{12}O_6$. And then the most abundant component of cells is water, from 80 to 90% in most plant cells.

The other 14 elements are needed by plants for various reasons, some as part of the structure of plants or their products, such as nitrogen in proteins and magnesium in chlorophyll, but others are essential in metabolic activities, in carrying energy, or in making enzymes function. A summary of the functions of these 14 mineral elements is found in the accompanying table.

Plants mainly use two forms of nitrogen, nitrate (NO_3^-) and ammonium (NH_4^+). These two ionic nitrogen carriers behave very differ-

Functions of Mineral Nutrients in Plants
(compiled from various sources)

NUTRIENT	PART OF ENZYMES	PROTEIN SYNTHESIS	PART OF PROTEINS	PHOTOSYNTHESIS	OTHER METABOLIC FUNCTIONS	CELL DIVISION	PART OF NUCLEIC ACIDS	PART OF CELL STRUCTURES	OTHER
N (nitrogen)		X	X		X		X	X chlorophyll	part of vitamins, purines, alkaloids, & metal complexes in enzyme functs.
P (phosphorus)	X			X	X		X	X membranes	part of metabolic energy carriers, vit. B_1; disease resistance
K (potassium)	X	X			X				water & pH balance; phloem transport; controls leaf movements & stomata
Ca (calcium)	X	X				X		X cell wall, membranes	
Mg (magnesium)	X	X (or Co)		X				X chlorophyll, ribosomes	aids stabilty of nucleic acids & membranes
S (sulfur)	X		X		X				
Fe (iron)	X	X		X	X				needed for chlorophyll synthesis
Cu (copper)	X			X	X				
Mn (manganese)	X			X	X				
Zn (zinc)	X								
Co (cobalt)	X	X (or Mg)							needed for N-fixing by bacteria
B (boron)					X	X			phloem transport
Mo (molybdenum)	X				X				needed for N-fixing by bacteria
Cl (chlorine)				X	X				water balance; part of alkaloids & indoles

ently in the soil, nitrate being easily leached into the groundwater, and ammonium being held on soil colloids and thus not easily lost. They also can have very different effects on plant growth. In many plants, fertilization by nitrate nitrogen produces rapid vegetative growth (stems and leaves); in fact, too much nitrate nitrogen can cause the plant to take up so much that it can cause nitrate poisoning in animals eating the crop. On the other hand, ammoniacal nitrogen promotes flower and seed or fruit development in many crops.

Only a relatively small amount of mineral nutrients is needed by plants. From 80-90% of the plant is water. Of the remaining 10-20% (the dry weight), carbon and oxygen make up nearly 90%. So only about 1-2% (some books say 5%) of a plant consists of mineral nutrients. But in spite of their small amount, they are absolutely necessary. If they are absent or deficient, the plant will die or be unhealthy. If a mineral deficiency is extreme, visible symptoms can be seen in the appearance of the leaves or other parts of the plant, but slight deficiencies are harder to detect. They may only be detectible as a reduced growth rate or sugar production rate, or as an amino acid imbalance, while the plant may *appear* to be green and healthy. At the opposite extreme, too much of some elements may be harmful or toxic. These include boron, sodium, chlorine, manganese, aluminum, iron, and copper.

Why fertilize? Since plants use mineral nutrients from the soil and are then harvested and not completely returned to the soil, and because some nutrients leach out of the soil or become chemically tied up and unavailable, the farmer must add some fertilizing materials to supplement or correct the soil's fertility and/or improve the soil's structure and pH. Other reasons for amending the soil in some cases are the correction of salinity and alkalinity (mostly in the western U.S.), helping soil to warm up faster in the spring, and overcoming waterlogging problems.

Types of fertilizer. We run into some difficulty when we try to classify fertilizers and soil amendments, because words have different meanings to different people. To a chemist, the word *organic* means any chemical substance containing carbon atoms, whether or not it is a natural or man-made chemical, and *inorganic* means non-carbon substances. But the original meanings of those terms were: substances produced by living organisms are organic and all other materials are inorganic. The latter meanings are useful to us here. Two other terms that can be confusing are *natural* (any substance found in nature, not produced by man) and *synthetic* (man-made, artificial). Some people often use the word *chemical* to refer to synthetic fertilizers, but this is not good because the term simply refers to any distinct kind of matter, whether natural or synthetic.

With regard to fertilizers and soil amendments, we may divide them into those derived from organisms, the organic materials, such as animal manures, sludge, bone meal, tankage, green manures, composts, and humic substances; and those which were not produced by organisms, the inorganic fertilizers. We can then divide the inorganics into those that are natural, such as mineral deposits (limestone, rock phosphate, potash, gypsum, greensand, granite, and fly ash); and those that are synthetic, including anhydrous ammonia, and the many "salt" fertilizers (chemically composed of a positive and a negative ion). We run into some difficulty here, because some salt fertilizers, including muriate of potash, are not man-made, but are mined from ancient salt deposits.

Organics. Still, these groupings are useful because their members share certain characteristics important to agriculture. Organic fertilizer materials contain a high proportion of complex substances (proteins, carbohydrates, fats, organic acids, vitamins, phenols, etc.), and when acted upon by living soil organisms, are slowly broken down to release *both simple and complex* plant nutrients; for contrary to the common belief of the "experts," plants not only absorb simple ions, but also can use more complex molecules, such as amino acids, amines, phenol, and indole. The nutrients in organic materials are very resistant to loss by leaching since they are either large molecules or are temporarily incorporated in the bodies of soil organisms. Organic substances generally provide plants with a variety of nutrients, not just one or two as in many inorganic materials. Organics will mainly be covered in the next chapter.

Inorganics. Most of the natural inorganic materials (rock deposits) have the advantage of breaking down slowly, and they usually supply plants with a number of trace elements in addition to their main constitutents. On the other hand, synthetic "salt" fertilizers are generally rapidly soluble (available to the plant) pure chemical substances (without trace elements). They quickly separate into their two ion components, some of which can be harmful to soil and soil life. Therefore, care must be taken in using the salt fertilizers to use those that cause the least harm and to use them properly, for if used in too great a concentration, salt damage can occur ("burning" of crops or seedlings), or the balance of soil nutrients can be disturbed.

Fertilizer application. There are many factors to take into consideration when applying fertilizers and soil amendments, and we cannot discuss all of them in detail here. But in general, we should consider the characteristics of the fertilizer material, the soil type and soil test results, topography of the land, climate, past fertilization and crop history, intended crop, use of herbicides and pesticides, time of application, liquid vs. dry fertilizer, cost, and whether you are on a fertility maintenance or building program. Leaving the organics for the next chapter, we will give some general principles for most effective use of inorganic fertilizers and soil amendments. They are all based on common sense, so if you understand some of the characteristics of fertilizer materials, how they function, how the plant functions, and the principles of ecology and the soil, you can figure out for yourself the best methods. Remember that the nutrients must be made *available to the plant*, in the *right amounts*, in the *right forms*, and at the *right time*.

Crop. High yield or high value crops will give more cost-effective use of fertilizers. Some crops respond to fertilization more than others (corn responds more than soybeans). Crops you intend to feed to your livestock should receive priority over those grown to sell. Different crops have different requirements: corn needs high nitrogen, while legumes do not; alfalfa needs high calcium.

Soil. Level, well drained loamy or clay soils, and soils with adequate organic matter can hold nutrients best and respond well to fertilization. Poorly drained, waterlogged, or drought-prone soils do not respond well. High sand content soils cannot hold nutrients well and require more fertilization. Very low (below 6.0) or high (above 7.5) pH reduces availability of several nutrients, especially phosphorus, except in very high organic matter soils, where pHs of 5.0-5.5 give best overall availability. Dry fertilizers will not be effective in dry soil. Fertilizer materials that cause a great change in pH should be avoided, or else applied long before planting or placed far from the seed.

Placement. Broadcasting can be done quickly and with low labor costs, and is good for a fertility building program, but for high-priced fertilizers, row application gives excellent results at a much lower cost. Also, row application does not nourish weeds between the rows. For row application, the fertilizer band should be placed below or to one side of the seed to prevent "burn" of seedlings. Leachable fertilizers should not be placed very deep; sidedressing of nitrate nitrogen is effective.

Timing. Since seedlings do not require much nutrients, a more efficient use of fertilizers can be made by using small split applications throughout the growing season ("spoon feeding"), especially of nutrients that leach readily, such as nitrate nitrogen. If a crop is too tall to drive an ordinary applicator through, a fogger, a "highboy" or an airplane with foliar application can be used, or fertilizer can be applied in irrigation water. Fall application of fertilizers that break down slowly (rock fertilizers, organic matter) is best for the next year's crop.

Salts. Synthetic fertilizers with a low "salt index" are preferred, in order to prevent salt damage to seedlings, roots, and soil life. The salt index is a measure of the fertilizer's ability to draw water out of living cells by osmosis. The salt index of some common fertilizers are (from *Farm Chemicals Handbook*, 1981, p. B74):

Material	*Salt Index*	*Material*	*Salt Index*
anhydrous ammonia	47.1	superphosphate	7.8
urea	75.4	triple superphosphate	10.1
ammonium nitrate	104.7	potassium chloride	116.3
ammonium sulfate	69.0	potassium sulfate	46.1
sodium nitrate	100.0	sulfate of potash-magnesia	43.2
potassium nitrate	73.6	calcitic limestone	4.7
calcium nitrate	52.5	dolomitic limestone	0.8
mono-ammonium phosphate	29.9	calcium sulfate	8.1
diammonium phosphate	34.2		

Note the much higher salt index of potassium chloride (muriate of potash), compared to potassium sulfate. This, besides the chlorine content of muriate of potash, is another good reason to use potassium sulfate (0-0-50) instead of muriate of potash.

Nutrient balance. Crops grow better and produce higher quality food if the soil nutrients are in a relative balance. The exact balance will differ slightly with different crops, but for most crops, the proportion of calcium should be higher than potassium. This will produce what is

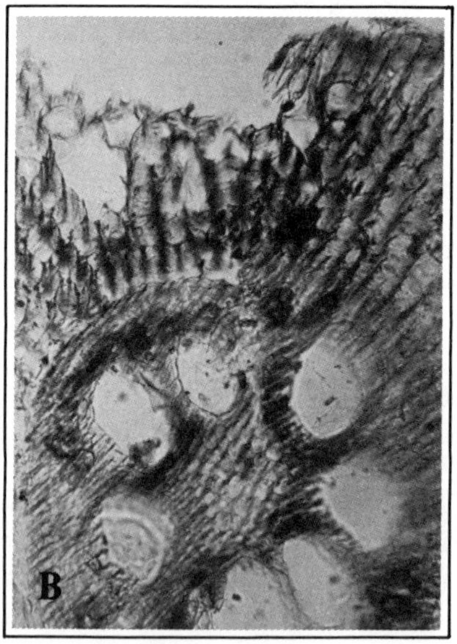

A. Corn roots grow *around* the zone where salt fertilizer was placed in the soil. B. Corn root tissues damaged by salt fertilizer (cross section, magnified 100 times under the microscope).

Nutrient Interactions

If the element (or soil condition) at the left is too high (H) or too low (L), a deficiency of the element at the top can result.

(information compiled from many books and journals)

	N	P	K	Ca	Mg	S	Fe	Al	Cu	B	Zn	Mo	Mn	Na	Cl	Co
N (nitrogen)			H							H	H					
P (phosphorus)	L		H	H			H		very H	H	L	H				
K (potassium)				H	H					H	very H					
Ca (calcium)		H			H		H			H						
Mg (magnesium)	H	H	H	H												
S (sulfur)	L											H	L			
Fe (iron)		H + Al + lowpH		H + Al,Mn + lowpH	H + Al,Mn + lowpH				H + Mn							
Al (aluminum)		H + Fe + lowpH		H + Fe,Mn + lowpH					H							
Cu (copper)							H					H				
B (boron)			H	L	H											
Zn (zinc)							H									
Mo (molybdenum)	L						H		H							
Mn (manganese)	H		H + Fe,Al + low pH	H + Fe,Mn + low pH			H + low pH		H + Fe							
Na (sodium)			H	H												
Cl (chlorine)			H													
Co (cobalt)																
Ni (nickel)							H									
pH		L + H Fe,Al		L + Al Fe,Mn	L + Al, Fe,Mn		H or L + Mn		H	H or L	H	L	H + org. m.			H
organic matter									H	L	H		H + H pH			

sometimes called proteinaceous crops, whereas too much potassium produces "carbonaceous" crops, high in carbohydrates and bulk (cellulose), but not nutritionally complete. Also, if phosphorus is low relative to calcium, animals or humans who eat the plants may suffer reproductive problems (from F. A. Gilbert, *Mineral Nutrition of Plants and Animals,* 1948). If there is too much magnesium relative to calcium, unhealthy plant cell walls may be formed. There is a host of other nutrient interactions that have been discovered; see the accompanying table.

Nutrient form. There are two main types of nitrogen used by plants, nitrate and ammonium. In some cases, nitrate nitrogen tends to produce lush vegetative growth, while ammonium nitrogen produces flowers and seed, if the plant is mature enough. So, if you are growing alfalfa for hay, use a nitrogen source that contains nitrate, but if you want grain, or seed alfalfa, use an ammonium-containing source.

Trace elements. Some people worry about trace element deficiences, and commercial fertilizers are available with trace element "packages." Trace element availability is affected by soil type, pH, and the balance of other nutrients, especially the major nutrients (see the nutrient interaction table). Except in cases where a severe trace element deficiency is obvious, it is best to first get the major nutrients in balance. Usually that will solve the trace element problem; if not, then it can be attacked. The trace element packages offered by fertilizer companies are just a shotgun approach and mostly a waste of money.

Dangers. To avoid or reduce stress on plants and harm to soil and soil organisms, the following materials should be avoided:

1. Ammonia or materials that release large amounts of ammonia (urea and diammonium phosphate). Large amounts of ammonia can kill soil organisms, destroy humus by making it soluble, and thereby cause soil hardening.

2. Chlorine-containing fertilizers (mainly muriate of potash, also calcium chloride and ammonium chloride). Although plants need *small* amounts of chlorine and most can tolerate a fairly large amount (up to 100-140 parts per million), and although chlorides can leach from the soil, amounts of chloride can temporarily be found in the soil that can not only injure plants, but can kill the beneficial soil organisms. Even if crops can tolerate chloride, their quality can still be lowered. High chloride is known to reduce yield and growth of corn. Potato and tobacco farmers usually avoid chlorine-bearing fertilizers, and the Chinese refuse to import any chlorine-bearing fertilizers.

3. Dolomitic limestone or other magnesium-containing fertilizers (such as sulfate of potash-magnesia or magnesium sulfate [epsom salts]) should not be used on soils that are not deficient in magnesium. The

magnesium causes nitrogen to be lost into the air, tends to form a crust on the soil, and causes low quality crops if the calcium level is low.

4. High concentrations of any salt fertilizer, especially those with a higher salt index, plus large applications of animal manures or whey, can cause salt damage to roots and kill soil life.

So we see that inorganic fertilizers and soil amendments are a "mixed bag" (no pun intended). Some are good, some are bad, and some must be used with caution. Now let's look at organic fertilizers.

CHAPTER 13

Waste or Resource?

There isn't much about manure to recommend it. It's sloppy, filthy, and it stinks. Most people would want to have as little to do with it as possible. But the wise farmer knows that this so-called waste material, and many other organic materials, are a potentially valuable resource which, if properly utilized, can greatly increase crop yields and quality.

Organic fertilizers. There are many types of material originating from living organisms that can be used as fertilizers and soil amendments. They include animal manures, green manures (fresh plants plowed under), seaweed and kelp, peat moss, tree bark, compost, leaf litter and leaf mold, grass clippings, sawdust and wood chips, brewery and cannery wastes, molasses residue, soybean and cottonseed meals, linseed and peanut meals, hulls and shells (of oats, rice, buckwheat, cocoa beans, and cottonseed), hay and straw, tobacco stems, castor pomace, papermill and sewage sludges, slaughterhouse wastes (bloodmeal, bone meal, hoof and horn meal, and tankage), fish meal, feathers, and humates (ancient organic deposits of salts of humic acids, somewhat like "petrified compost").

All of these organic materials are composed of mixtures of complex organic chemicals: carbohydrates, fats, proteins, amino acids, nucleic acids, vitamins, enzymes, aldehydes, alcohols, terpenes, organic acids, phenols, glucosides, esters, ketones, flavanoids, steroids, alkaloids, porphyrins, and so on—a veritable "soup" of thousands of different substances. These organic molecules are composed of atoms of the plant nutrients already familiar to us: carbon, hydrogen, oxygen, nitrogen, phosphorus, potassium, sulfur, magnesium, calcium, iron, etc.

Decomposition. When these complex molecules eventually break down into their simpler components, nutrients in the forms plants can use are made available to the plants. In the soil, organic matter is broken down (decomposed) by the soil organisms, primarily bacteria, actinomycetes, and fungi. You may want to review the details of decomposition in Chapter 6.

Humus. When organic matter decomposes under aerobic conditions (with air or oxygen), humus is produced. As we have mentioned in Chapters 3 and 6, humus is an extremely valuable substance in the soil since it stores plant nutrients and makes them available to the plant slowly, as the plant needs them. It also improves soil tilth, holds water, prevents erosion, and helps counteract and protect plants from excess salts, toxic chemicals, and excessive changes in pH. One of the constituents of humus, the humic acids, has been found to greatly stimulate plant root growth. The commercially available form, humates, is rich in humic acids.

The microorganisms that live in humus and help produce it, help in the above functions plus adding others: nitrogen fixation, producing a "glue" to hold soil particles together, producing plant growth hormones, preventing nutrients from leaching by temporarily making them part of their cells, protecting plants from diseases and pests, and breaking down (detoxifying) many man-made toxic chemicals.

Compost. The value of letting organic matter decompose before putting it in the soil was mentioned in Chapter 6. If you want to take the extra time and trouble it requires, composting can greatly benefit soil and crops. There are various methods of composting, and the books and articles on the subject will generally give very specific instructions to follow. Actually, satisfactory results can be obtained without being so fussy. Just remember a few general principles:

1. What you are trying to do is provide a home for a "volunteer army" of microorganisms. They need (a) food, (b) moisture, (c) air, (d) proper temperature, and (e) absence of toxic substances.

2. The fresh organic matter serves as food, although to be a good food source, it should contain nitrogen (mainly proteins) and considerable carbon (mainly carbohydrates: cellulose, hemicellulose, starches, sugars). A good carbon-to-nitrogen ratio is about 30:1. Materials with high cellulose content are slow to break down (such as straw, sawdust, and wood chips), so other materials should be mixed with them. Mixing animal manures with material of plant origin (straw, green manures, etc.) gives very good results in producing nutrient-rich compost. The recommended ratio in the famous Indore method of composting is one part animal manure, sewage sludge, garbage, or green manure to two parts cornstalks, straw, sawdust, or woodchips.

3. The organic matter should be piled or placed in such a way that a fairly large surface area is exposed. For agricultural applications, long windrows about 4-5 feet high and 8-12 feet wide at the bottom work well.

4. The material should be moist inside the pile, but not waterlogged. A moisture content of from 40-60% is ideal, but composting will occur from 25-75% moisture.

5. The pile or window should be turned one to several times during composting to help aerate it, mix it, or cool it. The more often it is turned, the more rapidly composting occurs, unless the temperature in the pile falls too low (below 85°F).

6. When other conditions are favorable, the temperature inside the pile will rise. Temperatures from about 100°-150°F are ideal. If the material gets hotter than that, it should be turned to cool it off.

7. Adding about 1/4 soil will help supply soil microorganisms to the pile, help absorb ammonia (conserve nitrogen), and speed composting. Adding a little lime or rock phosphate as a calcium source will stimulate microorganismic growth and prevent very acid conditions which kill bacteria. However the pH should not go much above 7, for that will also kill microbes. If you can smell a strong ammonia odor, the material is too alkaline.

8. Adding a special microbial "starter" is not necessary if the organic matter is already well supplied (as in animal manures), but it will speed up composting in other types of organic matter or in cold weather. Other good starters are some previously made compost, and compost "tea" made by soaking compost or rich soil in a container of water for several days.

9. Depending on weather (mostly temperature) and the other factors mentioned, it will take from 3-12 months for compost to be formed by using windrows. It takes longer in northern climates than southern. Special methods can produce compost more rapidly, in 4-6 weeks. Compost is "done" when the temperature inside the pile falls to near the outside temperature and no ammonia odor can be detected. Compost should be a dark, homogeneous material (no stems or leaves remaining) with a rich "earthy" smell (no unpleasant odors). The carbon to nitrogen ratio should be about 10 or 12:1 (this cannot easily be measured without a laboratory analysis).

Compost is essentially humus and can be put on soil at any time without harming soil or crops. It is like instant plant food. It is best to till it into the soil to keep its bacterial population active. Good results have been obtained at application rates from 1/2 to 10 tons per acre.

Direct soil application. The other approach to using organic fertilizers is to put them directly on or in the soil rather than composting them first. This has the advantages of requiring less time and trouble, and in conserving some nutrients which escape into the air during composting (nitrogen and carbon dioxide), but the disadvantage or danger of possibly damaging a crop if done in the wrong way.

Since decomposition of fresh organic matter temporarily ties up nutrients in the cells of microorganisms, large amounts of fresh organic matter should not be plowed or injected deeply into the soil (below 5-7 inches) or it may decompose anaerobically (putrify) and produce toxic substances.

Large amounts of animal manures can cause salt damage and an over supply of potassium, if the manure cannot decompose aerobically. Anaerobic putrification can easily occur when large amounts are applied.

Manure management. For those farms with livestock or poultry, manure is a natural by-product, a "free" source of organic matter which should be used as a valuable source of nutrients. Unfortunately, many farmers do not make full use of its potential, or mis-use it with harmful results, and some even make no use of it at all.

Like other organic matter, fresh manure contains nitrogen, phosphorus, potassium, and other elements that *could* be used by crops if the manure is handled and applied in ways that conserve its nutrient value. But by using wrong management methods, up to 80% of the nitrogen can be lost—or even nearly all of the nutrients—or else toxic salts and other toxic decomposition products can damage crops. Again, by using certain common-sense principles and by understanding soil ecology, we can avoid the pitfalls and reap the benefits of using manure.

There is a variety of manure handling, storage, and application methods, ranging from daily spreading to stacking of solid manure to storage of liquid manure slurry, later spraying it on or knifing it into the soil. These general manure management principles can be applied to any system:

1. As with other organic matter, manure must decompose aerobically to form humus in order to benefit crops. It should be worked into the aerobic zone of the soil (upper several inches) where conditions are favorable for aerobic microorganisms.
2. Thinner applications over more acres is more effective than too large an amount, which can smother plants and cause salt damage, nitrate poisoning in animals, and pollution of groundwater, or become anaerobic and produce toxic substances.
3. Apply most manure in the fall, so it has time to decompose before crop growth and does not compete with crops for nutrients.
4. Save the liquid portion (urine), since it contains valuable nutrients.
5. The use of bedding increases the value of manure and also absorbs urine.
6. Avoid or reduce nutrient loss by:
   a. Working manure into the soil as soon as possible after spreading.
   b. Spread on cool, rainy days if possible.

c. Do not spread on snow or frozen ground, especially on sloping ground.
d. Do not stack manure where leaching or loss of urine can occur. A concrete floor is helpful.
e. Additives can be used to tie up ammonia, such as superphosphate (0-20-0); they should be added to manure as soon as possible and mixed as thoroughly as possible.
f. Composting to temporarily tie up nutrients in microbial cells and humus.

7. If additional materials are added to increase manure's fertilizer value (green manures, fertilizer supplements, lime, rock phosphate), mix them is as thoroughly as possible. Tests have shown that such supplements greatly increase the nutrient value of manure, more so than applying the materials separately (E. M. Bodrova & Z. D. Ozolina, *Simultaneous Application of Organic and Mineral Fertilizers*, 1965).

8. "Hot" manures (poultry, pig, horse) should be used with care to avoid plant injury. They can be mixed with cow manure, composted, or applied in small amounts well before crop growth.

Up to Date?

From *The Farmer's Every-day Book*, by J. L. Blake, 1851:

How to Preserve Manure–"If you cannot use all your manure, place it in heaps and cover it with earth two feet in thickness, which will inhale and retain most of its enriching gases till wanted. Nevertheless, it is better every year to put on your land all the manure you possibly can scrape together. Leave not a particle in your barnyard. All you can get from it before another season is clear gain, for it will lose but little more underground with a crop over it, than exposed to the action of the sun and atmosphere."

Making Manure–"A correspondent of the Maine Farmer says that he converted straw, corn-stalks, and potato-tops into good manure in fourteen days, only by heaping them together, and mixing unslaked lime with them. He used six casks of lime, and had fifty loads of good manure. The loads were such as farmers usually carry, a large half cord to the load."

How to Get Manure Cheap–"As soon as the summer manure is scraped out of the cow-yard, spread over it again plentifully, leaves, litter, loam, peat-mud, or almost anything else. Fail not to neglect it. The cost to do it will be much less than to purchase guano or poudrette, and it will be more useful, for you will be able to improve the texture of your soil, if you exercise good judgement in selecting your materials."

9. Rates of fresh manure application depend on soil conditions, crop to be grown, time of year, and kind of manure. Up to 10 tons per acre of solid manure is generally safe (2500 gallons per acre of liquid slurry), but less may give better results. "Hot" manures should be applied at lower rates.

Organic vs. synthetic. Organic fertilizers have a considerable number of advantages compared to synthetic "salt" fertilizers (see Chapters 11-12). Of course, some salt fertilizers are worse than others, and if properly used, the safer ones can give good results. But in general, the use of synthetic fertilizers *alone* can lead to a number of problems:

1. Can be toxic to plants and soil life (salt damage, chlorine), or animals that eat the crops (nitrate poisoning).
2. Can destroy humus (excess ammonia).
3. Can destroy soil structure (tilth) by destroying humus or causing crusting or hardpan.
4. Can pollute surface and groundwater (nitrates, phosphates).
5. Can upset nutrient and pH balance in soil; provide only a few nutrients; plants are unhealthy and vulnerable to diseases and pests; crops are of low quality.
6. Provide only a quick "shot in the arm" of some nutrients (not long, slow release).

If organic fertilizers are used improperly, they also can cause problems, such as pollution and salt damage from animal manures, but on the whole, organic fertilizers and natural inorganic minerals offer the following advantages:

1. Are safe for plants and soil life; stimulate soil life.
2. Build up humus.
3. Improve soil tilth.
4. Hold nutrients, do not pollute water.
5. Feed plants a "balanced diet;" often provide trace elements and growth-promoting substances; produce healthy plants that are resistant to diseases and pests, and high quality crops. Increase soil's buffering capacity to resist pH change and absorb toxic substances or detoxify them.
6. Release nutrients slowly, in greater amounts in warmest months, when plants need more nutrients.

But one should not conclude from the above lists that synthetic fertilizers are totally bad and that organics and natural inorganics are totally good. Anything good can be mis-used. Some synthetics are definitely harmful and should be avoided (see Chapter 12), but others have their place. However, synthetic salt fertilizers should not be used alone. Any farming operation should always include the use of some organic and natural inorganic fertilizers and soil amendments. Tests have proved that mixing rock fertilizers or synthetics with manures gives much better results than using either separately, or even together but not mixed. Experiments in Russia found that potatoes grown on sandy loam soil yielded 7,930 lbs/A without fertilizer (12 year average); 15,325 lbs/A with 16 tons of manure per acre; 16,483.5 lbs/A with mineral fertilizers corresponding to the N,P,K,Ca content of 16 tons of manure per acre; and 18.265.5 lbs/A with 8 tons of manure per acre *plus* mineral fertilizers corresponding to the N,P,K,Ca content of 8 tons of manure per acre. (E. M. Bodrova & Z. D. Ozolina, *Simultaneous Application of Organic and Mineral Fertilizers*, 1965).

Proper fertilization is great—a must, but it's not the only thing. There's more to the story.

CHAPTER 14

Scratching the Surface

Proper use of fertilizer is certainly important, but tillage methods and cropping systems—what we do to the soil, and when—play a large role in producing optimum yields. An additional major consideration *should be*: what *long term effect* does the cropping system or tillage method have on the soil and soil life? No conscientious farmer can afford to do anything that will eventually ruin the soil, even though he may get fantastic yields (for a while). Unfortunately, not all (very few, actually) of the textbooks, agronomists, and extension agents look that far down the road. They let bushels and dollars take precedence over organic matter and soil life. In the well known textbook, *Soil Fertility and Fertilizers* by S. L. Tisdale and W. L. Nelson (2nd ed., 1966), we read: "Maintenance of organic matter for the sake of maintenance alone is not a practical approach to farming" (p. 564); and in a section on whether legumes should be used to supply nitrogen, or commercial fertilizers, we read: "The selection of the program that the farmer should follow becomes a matter of economics; he should choose the program yielding the greatest net return on his investment" (p. 566).

Proper fertilization and tillage methods can give terrific yields AND not harm the soil, AND produce *high quality* crops at the same time—something that "traditional" agricultural methods almost never can do. Let's look at some of the commonly used cropping systems and tillage methods. Many factors should be considered: soil type and condition (compaction, hardpan, temperature, herbicide carryover, soil erosion, slope and drainage, subsoil), crops to be grown and their nutrient requirements, root depth, organic matter returned to the soil (plant residues, manures), soil fertility and nutrient balance (soil test results), weeds and pests, planting depth and time, plant population and row width, tillage and seedbed preparation, and markets.

Is rotation necessary? For decades farmers have been advised to rotate their crops to prevent depleting their soil, and now we often hear the "experts" tell us not to rotate. Which is best?

There are advantages and disadvantages to either system, and it depends on your local situation and needs. First of all, if the soil is properly fertilized and if organic matter and soil life are fostered, it is not necessary to rotate. Each crop species has its own set of nutrient requirements and soil type preferences, and if you get your soil "fine tuned" for one crop, and if it pays to keep raising it (if market prices hold up, or if you need it to feed livestock), you can raise the same crop year after year. As we will see in Chapter 16, even weeds and insects do not have to be a problem in properly fertilized soil.

Rotations usually include a legume crop to add nitrogen to the soil and a variety of deep and shallow-rooted crops with the intention of improving tilth and utilizing nutrients from different depths. These are good reasons and they do have their value, but, if you don't want to or need to, you don't have to rotate. A possible exception is reported in *Soil Conditions and Plant Growth,* 9th ed., by E. W. Russell (1961): continuous clover causes an accumulation of a toxic substance secreted by clover roots which will inhibit plant growth; however it can be counteracted by adding large amounts of manure.

In tests done at the University of Missouri Agricultural Experiment Station, growing a corn-wheat-clover rotation from 1917-1941 on soil which had previously been cropped for 60 years did not increase the amount of nitrogen significantly (0.136% in 1917 vs. 0.137% in 1941), while a continuous bluegrass sod increased nitrogen from 0.105% in 1917 to 0.132% in 1941, and red clover turned under increased nitrogen from 0.137% in 1917 to 0.170% in 1941. The latter two crops would allow humus to build up in the soil, while the rotation did not. Some rotations are definitely harmful. According to University of Minnesota soil scientist G. R. Blake, a series of row crops (which require a lot of tillage) can deplete the soil, since not enough plant residues are returned and too much tillage destroys soil structure (*Crops & Soils Magazine*, March 1980, p. 10-11).

Continuous cropping is said to encourage the build up of soil-harbored diseases and pests. Continuous corn has been grown at the University of Illinois test plots since 1876 with no indication of an increase of insects or diseases. Continuous cropping can destroy soil tilth and encourage erosion (if little organic matter is recycled or if soil organisms are destroyed), can cause more soil compaction from increased cultivation and harvesting traffic, and can deplete organic matter from increased cultivation. These dangers are minimized or overcome by a "living" soil with plenty of humus. With plenty of humus and good soil tilth, and by use of certain tillage methods to break up a hardpan, roots will grow deeper and utilize subsoil nutrients and be better able to withstand drought.

Tests Run at the North Carolina Agricultural Experiment Station Supplementing Corn with Varying Amounts of Nitrogen.

Nitrogen (lbs/A)	*Grain (bu/A)*	*Cornstalks (lbs/A)*
0	22.8	3,614
40	52.1	3,356
80	81.7	5,078
120	100.5	5,459
160	102.3	5,328

Continuous cropping has the advantage of allowing forage crops to be grown on sloping land to reduce erosion and row crops to be grown on level land. These two types of crops require a different balance of soil nutrients (alfalfa requires high calcium), so the soil can be fertilized for one crop, not several. If you want to specialize on only one crop, you can, and you will need less machinery than for several crops.

On the other hand, rotation has advantages. It does improve tilth and if properly done can build up soil nutrients. It uses natural crop plants to do these things rather than commercial fertilizers. It will take longer to get results. Also, tests have shown that supplementing crop residues with commercial fertilizers gives increased yields (just so you are careful to fertilize properly to keep nutrients balanced and quality high) and increases the amount of stalks (plant residues) that could be recycled to form more humus.

So whether you want to rotate crops or continuous crop, is up to you. Either cropping system can have advantages, IF you use proper methods and do not deplete soil fertility, humus, and organisms.

To till or not to till? In recent years, no other farming method has been so ballyhooed as reduced tillage systems, whether called minimum-till, conservation-till, or no-till. It is promoted by farm papers and magazines, fertilizer and pesticide literature, universities and their extension agencies, the USDA, and even ordinary newspapers. Local farmers' testimonials are usually included. The Wisconsin Agricultural Stabilization and Conservation Service is "promoting conservation tillage with grants that pay as much as $35 per acre to farmers who switch from traditional methods" (article in the Milwaukee Sentinel, Mar. 1, 1982). The advertising has been working. The number of acres put in conservation tillage has grown from less than 5 million in 1963 to over 55 million in 1979.

The two big problems which reduced tillage is promoted to overcome are (1) soil loss by wind and water erosion, and (2) energy—fuel and labor costs. Additional advantages are increased water absorption (and decreased evaporation), being able to farm more sloping land (by reducing the erosion danger), being able to plant and harvest under a wider range of soil moisture, and needing less machinery (no plows, no large tractor, no cultivator, etc.). Basically, reduced tillage systems leave a considerable portion of plant residues on the surface and use a minimal number of passes over the field. The plant residues act as a mulch and reduce wind and water erosion, and aid in water absorption. Water erosion can be reduced by as much as 50-fold, according to a study in Ohio. Fewer passes over the field obviously save fuel and labor. All of these advantages exist. In a study of no-tillage agriculture by several University of Kentucky professors (reported in *Science*, June 6, 1980, p. 1108-1113), the total energy required for no-till versus conventional tillage methods, from machinery needed to fertilizers to seed to herbicides and pesticides to planting to harvesting was calculated. They found that no-till methods would use 4.39 million Calories of energy compared to 4.72 million Calories for conventional tillage. The main saving in energy of no-till is in planting costs. A University of Nebraska study showed that conventionally tilled corn required 5.33 gallons of diesel fuel to plant, while no-till corn only needed 0.90 gallon. Sounds great, doesn't it? But let's look at the rest of the story.

There are many disadvantages of reduced tillage and many problems *it creates*. First, a farmer can't just go out and use conventional tillage machinery and methods in a reduced till fashion. Conventional planters and drills generally do not work well with plant residue covering the surface, so you have to buy a new planter. The plant residues form a protective overwintering home for insect pests and plant diseases, so more pesticides are required (about 50% more for corn). The surface mulch reduces the soil temperature in the spring (as much as 10.8°F lower at 1 inch depth), so you need a cold-resistant variety if you live in the central and northern U.S. The lack of tillage encourages weeds, especially grassy ones, so more herbicides are required. Fertilizers and soil amendments have to be applied on the surface, so more fertilizer must be used to let enough work its way down to the roots (some planters can place fertilizer deeply in the row). Recommendations now are to split the application of nitrogen, with some applied with the main fertilizer application at planting and the rest applied as sidedressing later; this means an extra trip over the field! Yields in some cases are lower with reduced tillage (for corn on poorly drained soils). By keeping most of the plant residues on the surface, most of the potential nutrient value from their decomposition is lost by oxidation-mineralization (see Chapter 6), and they do not contribute to soil tilth. No-till methods do not work well if the soil pH is below 6.6, but most crops do best from about 6.0-6.8.

With all those disadvantages, why are reduced tillage methods promoted so heavily? The necessary and increased use of herbicides and pesticides, and the higher amounts of fertilizer needed may explain why the fertilizer and chemical companies are pushing reduced tillage. And remember in Chapter 7, the close link we found between industry and the university? If we consider the long term effects of poisoning the soil with herbicides and insecticides, killing soil life, and depleting the humus, whatever dollars are saved in the short term by reduced tillage just aren't worth it!

A better way. Wouldn't it be great to have most of the advantages of reduced tillage and none of the disadvantages? You can—by using the natural systems approach to agriculture presented in this book, plus certain tillage methods. As we have mentioned in Chapters 3 and 6, a soil with plenty of humus and soil organisms will resist water and wind erosion (remember the "glue" secreted by the microorganisms?) and will be spongy and will soak up and hold a lot of water. Disking plant residues under, with a small amount sticking out to the surface will aid water absorption by a wicking action, plus the below-ground organic matter will decompose into nutrient-rich humus. Soil with good tilth will be easier to work, resulting in some fuel savings and obviating the need for large tractors. Humus-rich "living" soil will fight soil insect and nematode pests and plant diseases. Soil rich in organic matter warms up faster in the spring; also the use of ammonium sulfate helps warm up soil. Ridge planting, where the soil is formed into a ridge every 40 inches, is an effective tillage method which helps conserve moisture and helps warm up the seedbed on the tops of the ridges (in drought-prone areas, the seeds should be planted between the ridges in the "valleys").

It is wise to take the temperature of your soil in the spring, for some seeds (corn and soybeans) will not germinate until the temperature at planting depth reaches about 55°F, or better yet 60°. Take the temperature in the morning, because direct sunlight can raise the temperature 8 or 10°F.

In row crops, high plant populations and narrow rows may or may not pay off. More plant nutrients will be needed, so the soil must be at peak fertility. Some hybrid varieties produce higher yields in higher populations and some do not. In dry years, high plant populations release considerable soil moisture by transpiration through the leaves.

Hardpan. To farm more of your soil, break up a hardpan by plowing an inch or two deeper at a time or use a chisel plow or subsoiler. Plowing under lime will help improve the soil tilth and overcome a hardpan. Even deep-rooted plants such as alfalfa cannot effectively penetrate a dense hardpan. Soils have a large store of nutrients "on reserve" in the subsoil, so why spend extra dollars for fertilizers when you could use what already is there? A hardpan also severely limits water absorption, leading to scenic-looking lakes in wet weather.

Weeds. Weeds can in part be controlled by two tillage methods (see Chapter 16 for other weed control methods). Weeds can often be controlled by making sure the crop gets out of the ground first and outcompetes and shades the weeds. This can be done by preparing an optimum seedbed *in the row*, but leaving the soil between the rows in poor condition. The other method of weed control by tillage is the familiar use of cultivation during the growing season, being careful not to cut the roots of the crop plants. Cultivation has the advantages of not only killing weeds, but also of breaking a surface crust and helping aerate the soil. However, too frequent cultivation tends to destroy humus by causing too rapid microbial decomposition.

Plowing. Fall plowing is preferred over spring plowing because freezing and thawing help improve tilth, plus hibernating insect pests are more likely to be killed from exposure.

For ordinary tillage and seedbed preparation, deep moldboard plowing is not recommended. It turns the soil completely over, covering the former aerobic top-most soil and possibly turning it anaerobic, and it brings up relatively lifeless anaerobic soil. The whole soil "ecosystem" is disturbed, and it takes time to readjust. It is better to chisel plow or disk. Tilling too-wet soil causes clod formation and compaction from machinery. Occasional plowing is useful in good, aerobic soil, since it increases the depth of aeration and works humus down to enrich the subsoil. Shallow moldboard plowing (just below the aerobic zone) in the fall is a beneficial practice. The depth of the aerobic zone can generally be determined by noting how far down the small feeder roots and root hairs are on a plant you have dug out of the ground.

Of course, on sloping land such common-sense conservation practices as contour plowing, grass waterways, and strip cropping are beneficial, although one should ask himself if planting row crops on steeply sloping land is wise land use. To reduce erosion, forage or pasture usage should be followed instead.

Multiple cropping. One cropping system which offers the promise of greatly increasing the productivity of an acre of land is multiple cropping, growing two or more crops in one growing season. Multiple cropping has long been practiced in the southern U.S., but recently has been used more in the North. It is promoted along with reduced tillage systems, since as much time as possible should be saved, and plowing and seedbed preparation are eliminated in reduced-till methods. We have already covered reduced tillage; multiple cropping can be done just as well using conventional tillage methods, provided one is careful to maintain soil fertility and tilth, and provided the soil's water supply is adequate.

Depending on the latitude (climate), various crop sequences can be used. In the South, winter wheat-soybeans works well, as does wheat or barley followed by corn for silage, or small grains for silage followed by corn for grain. In some cases, even as far north as Wisconsin, it is possible to double or triple-crop by growing one or two crops of vegetables, which have a short maturity time, followed by a winter cover crop, such as rye. Often, one crop can be aerial-seeded into the preceding one while it is still growing, such as seeding wheat into standing soybeans.

CHAPTER 15

Is "'Organic'" the Magic Word?

By now, you probably realize that we do not advocate the use of toxic agricultural chemicals that kill soil organisms and pollute the environment, and that we recommend building up the organic matter and humus content of the soil. Does that make us "organic farmers"?

What is "organic" farming? J. I. Rodale's *Encyclopedia of Organic Gardening* defines it: "Organic gardening or farming is a system whereby a fertile soil is maintained by applying Nature's own law of replenishing it—that is, the addition and preservation of humus, the use of organic matter instead of chemical fertilizers, and, of course, the making of compost and mulching."

Organic farming and gardening can be described as a movement and a philosophy, which has a growing number of "converts." A strict adherent to organic methods absolutely refuses to use any "chemical" fertilizers, or pesticides or herbicides, and will only use "natural" fertilizers, soil amendments, and pest, weed, and disease control practices.

The pioneer of the modern organic movement was Sir Albert Howard, a British horticulturist who was director of the Institute of Plant Industry at Indore, India. He developed a method of composting agricultural and city "wastes" called the Indore method. This is described in his book, *An Agricultural Testament* (1943) (He first published his Indore method in an earlier book, 1931). Since then, disciples of the organic movement have "spread the word" and have made many "converts" on all continents. And with good reason, for organic agriculture is based on ecologically sound principles. It seeks to obey and work with the laws of nature, not to modify and violate them. Many things organic agriculture followers do is good—but do they do enough?

Yes, but . . . Many organic farmers and gardeners have a basic mistrust of anything "scientific," associating science with "chemical" fertilizers and pesticides. Consequently they often do not have their soil tested and do not have a basic understanding of soil ecology, soil chemis-

try, and plant functions and nutrition. They assume that using "natural" fertilizers and soil amendments will automatically supply all of the plant's major, secondary, and micronutrient needs. Now, completely organic methods MAY supply all of the plant's needs, but then again they may not. How can you tell unless you know what the needs are and unless you test your soil frequently to see what is going on? An organic farmer may happen to be doing everything just right and may have fantastic yields of high quality crops . . . BUT, some organic farmers have out-of-balance soil and poor quality, moderate-yielding crops. Many organic farmers get good results the first several years they switch to organic methods, but then their yields and quality gradually decline and pest problems increase.

If we review what we learned about decomposition in Chapter 6, we recall that when soil microorganisms decompose organic matter, plant nutrients are temporarily "tied up" in the microbe's cells, and the nutrients only become available to the plants *slowly*. Also, microbial activity can be slowed or stopped by the wrong temperature or pH and lack of oxygen or moisture. Natural rock fertilizers and soil amendments (lime, rock phosphate, greensand, etc.) also break down slowly. So unless conditions are ideal, a crop may not receive enough nutrients, especially during the critical maturing period. This is very common in these days of high yielding hybrids and high plant populations. Thus, in some cases, organic matter alone can't do it all.

What we must do then, is use organic methods as much as possible, but SUPPLEMENT the organic matter and humus with certain synthetic fertilizers, *if necessary* (using frequent soil tests to monitor conditions). But only those synthetic fertilizers that do not harm soil or soil life, and only in "safe" amounts.

Five-year tests run at the Vermont Agricultural Experiment Station by F. R. Magdoff showed that manure application to corn should provide all of the nitrogen needed by the crop, but that in finer textured soils (clays) and in wet years, not enough nitrogen can be provided by ordinary manure application rates, and supplementary inorganic nitrogen (ammonium nitrate) is effective (*Agronomy Journal*, July-August 1978, p. 629-632). Also, don't forget that whenever part of a crop is harvested, removed from the land and not recycled, some nutrients are removed from the soil and not replaced. If a deficiency develops (it may not; frequent soil tests should tell), supplementary fertilization above and beyond organic matter may be necessary.

Note the results of the North Carolina tests of supplementary nitrogen on corn yields (box in Chapter 14). Similar results are obtained by supplementing other nutrients. Of course, university tests usually are

TESTS RUN AT THE NORTH CAROLINA AGRICULTURAL EXPERIMENT STATION SUPPLEMENTING OATS WITH VARYING AMOUNTS OF PHOSPHORUS.

phosphorus (lbs/acre)	*grain (bu/acre)*	*straw (lbs/acre)*
0	15.2	1,108
4.4	86.7	5,730
6.2	85.3	5,537
11.0	121.6	6,542
22.0	126.3	10,733
33.0	127.5	8,246

not run under strict organic/no-toxic-chemical conditions, and scientists are concerned only about yield, not quality. We will discuss how quality can be measured in Chapter. 17.

Natural vs. synthetic. Most standard textbooks, professors, extension agents, and fertilizer salesmen say that there is no different between crops grown by organic methods and "chemical" methods—that the nutrients used by plants are the same inorganic ions, no matter what source they came from. That the food value of the crops is the same—no difference in vitamin content, proteins, etc. That vitamins derived from living organisms are identical to man-made vitamins. This argument seems true to a chemist and is reassuring to those involved in the synthetic fertilizer and vitamin business—*but* "it ain't necessarily so."

Some things the chemist doesn't measure are flavor and texture. Vegetables raised by N P K technology on "dead" soil are tough, fibrous, and tasteless, while those grown on humus-rich soil are tender and flavorful (Sir Albert Howard, *An Agricultural Testament,* 1943, p. 82; E. B. Balfour, *The Living Soil*, 1975, p. 141, 144). Plants grown in manure-fertilized soil contain more vitamins than those grown with other fertilizers (E. B. Balfour, *The Living Soil*, 1975, p. 128, 133; J. I. Rodale, *Pay Dirt*, 1950, p. 70). Both humans and animals prefer compost-grown food, and a smaller quantity of such food is needed to produce a given weight gain (E. B. Balfour, *The Living Soil*, 1975, p. 141). Dr. William Albrecht experimented at the University of Missouri by fertilizing timothy hay with extra synthetic nitrogen. The extra nitrogen increased the "crude protein" content of the hay (crude protein is found by multiplying the amount of nitrogen found by chemical analysis by 6.25). But when the hay was fed to weanling rabbits, 70% of them died during a summer heat wave. When the experiment was repeated and the rabbits were switched to red clover hay grown without extra nitrogen, no more deaths occurred. Although the two hays had nearly equal amounts of *total* amino acids, the *balance* of amino acids was different (*The Albrecht Papers*, 1975, p. 171-173). The excessive salt fertilizer apparently upset the plant's metabolism and caused it to produce a balance of amino acids which was not conducive to good health in animals. It has been found that excessive levels of nitrates and nitrites in livestock (at subtoxic levels) cause depressed thyroid gland activity, leading to reduced conversion of carotene to vitamin A, leading to reduced ability to synthesize RNA (ribonucleic acid), which is necessary for normal cell metabolism (*Unfit for Human Consumption*, by Ruth M. Harmer, 1971). Dr. Barry Commoner of Washington University, St. Louis, has found that synthetic nitrogen compounds have a different isotope makeup than natural nitrogen compounds. Plants grown with synthetic nitrogen fail to field ripen and are readily attacked by insects (*An Acres U.S.A. Primer*, 1979, p. 152).

Is it really organic? Some fertilizer companies proudly advertise what they call "organic" fertilizers, and many organic farmers buy them. But the truth is, that most commercial so-called organic fertilizers are mixtures of synthetic inorganic salt fertilizer materials, including muriate of potash (potassium chloride). How can this be? The fertilizer companies get away with this because the fertilizer laws allow any fertilizer containing a certain percent of organic nitrogen material to be labeled and sold as "organic." However, the usual source of "organic" nitrogen is the totally synthetic material, urea. Urea is technically called an organic substance by chemists because it contains carbon (its chemical formula is $CO(NH_2)_2$). So the fertilizer company puts in the minimum legally required amount of "organic" nitrogen and fills up the rest of the ton with possibly damaging synthetic inorganic materials, almost always including muriate of potash, which is nearly half chlorine.

As we have cautioned in Chapters 11 and 12, you need to be very careful and know what source materials went into any fertilizers and soil amendments you put on your land. It could make the difference between success and failure.

The future? Strictly organic agriculture and complete reliance on synthetic fertilizers and pesticides are at opposite ends of the agricultural spectrum. Although initially, 30 years ago, "chemical" farming seemed to be a messiah, conditions now clearly show that it has no future (or if it is continued, mankind has no future!). The environmental damage and sickness it causes are just too great. Crop and animal quality are sinking lower. Pest and weed problems are increasing, not disappearing. Economically, it cannot at all be defended. It uses non-renewable and in-

Organic Farming Gains Respectability

In 1980, the U.S. Department of Agriculture published a 94 page summary report of a one-year study of organic farming in the U.S. and Europe, titled *Report and Recommendations on Organic Farming*. The 10-member study team investigated the nature and activity of organic farming, the motivations of organic farmers, the level of success and economics of organic farming, and identified research and education programs to benefit organic farmers.

The study team found that there is increasing concern in this country about "the adverse effects of our U.S. agricultural production system, particularly in regard to the intensive and continuous production of cash grains and the extensive and sometimes excessive use of agricultural chemicals."

They found that most organic farmers were highly experienced, well educated, and had shifted to organic farming after a number of years of conventional (chemical-intensive) farming. Reasons to change to organic farming included soil health, food safety, environmental protection, soil and water conservation, and superior food quality. Organic farmers are generally good managers who are dedicated to responsible husbandry of soil, crops, and livestock; they place a very high value on environmental quality. They have not "regressed to agriculture as it was practiced in the 1930's. While they attempt to avoid or restrict the use of chemical fertilizers and pesticides, organic farmers still use modern farm machinery, recommended crop varieties, certified seed, sound methods of organic waste management, and recommended soil and water conservation practices."

The study team was "impressed by the ability of organic farmers to control weeds in crops such as corn, soybeans, and cereals without the use (or with only minimal use) of herbicides."

Crop yields on organic farms averaged comparable to conventional

continued. . .

creasingly expensive resources, oil and natural gas, as its major input and gives us devastating ecological and health damage as by-products. The sun is setting for totally synthetic-based agriculture.

Organically oriented agriculture, or the natural-systems approach which we promote, relies on the established *natural systems* and *cycles* around which the natural world is built. Nutrients are *recycled,* not made from oil by chemical plants. The beneficial soil organisms are fostered to help make nutrients available to plants, to fight diseases and pests, and to keep soil in good condition. Only natural-systems agriculture really makes sense.

But blind reliance on total non-synthetic chemical ("organic") methods, while much better than drenching the landscape with toxic

Organic Farming Gains Respectability, continued

farms; some crops (corn, wheat, and potatoes) gave lower yields on organic farms, but others (alfalfa, soybeans, and oats) gave the same or higher yields. Yields on organic farms tended to be higher than conventional farms in dry years, comparable in normal years, and lower in wet years. It generally took several years to make the transition from conventional farming to high-yielding organic farming.

Although organic farming requires higher labor and fuel costs than conventional farming, the savings in fertilizers and chemicals means that organic farming uses less energy than conventional farming.

The study team believes that organic agriculture is best suited for small farms which have enough livestock to supply adequate manure to maintain soil organic matter. If all U.S. farms with annual sales of less than $20,000 switched to organic farming, there would be a significant economic impact, mainly from market adjustments to decreased production of some crops (corn and soybeans) and increased production of others (hay and oats). Domestic needs could be met, but export needs of corn and soybeans would fall short. A computer study of the impact of *all* U.S. farms switching to organic agriculture was made. It was concluded that there would be a "major economic impact," but that such a shift could be made with less impact in a 3 to 5 year transition period.

But a more recent syndicated newspaper article by Ward Sinclair (June 23, 1982) does not sound optimistic. Apparently, the secretary of the U.S. Department of Agriculture and top department officials are "doing their utmost" to keep organic farming from receiving research funds. "There are, of course, big bucks business implications in the fight over organics. The multibillion-dollar fertilizer industry, as an example, could suffer seriously if farmers moved to organics in a massive way. Ditto other chemical companies whose products are vital to most U.S. farmers today."

substances, can be incomplete and can be producing poor quality crops from out-of-balance soil. A fear of anything "scientific" is like sticking your head in the sand. We need to use whatever is good from modern science and technology. The tricky part is sifting out the good from the bad. This is where a sound store of knowledge about natural systems and a lot of common sense are needed. If limited use of certain synthetic fertilizer materials, along with promoting humus formation and soil life, produces superior results and does not harm natural systems, then why not follow that method?

The totally organic farmer is certainly not "out of it" by not using the latest pesticides, nor is he a wild-eyed radical. He is on the right track and is far ahead of most of his "progressive" neighbors.

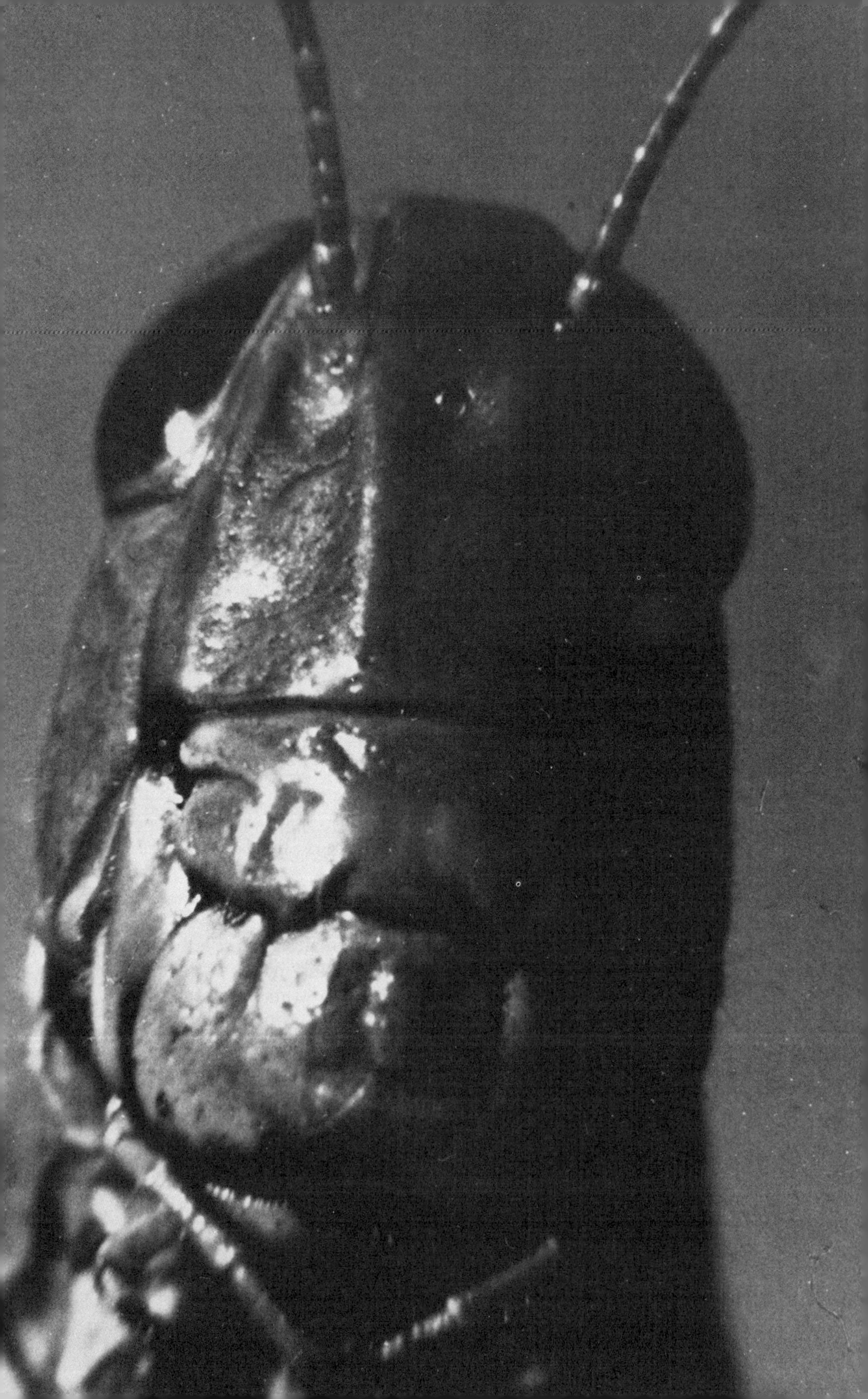

CHAPTER 16

Shoo!—
Pest and Weed Control

> "DANGER! May be fatal if swallowed, inhaled or absorbed through the skin. Rapidly absorbed through the skin. Repeated inhalation or skin contact may, without symptoms, progressively increase susceptibility to poisoning. Do not get in eyes, on skin, on clothing. Do not breathe dust. Do not contaminate food or feed products. Keep out of reach of children. Keep out of reach of domestic animals. Not for use or storage in or around the home. This product is toxic to fish, birds and other wildlife. Keep out of any body of water. Do not apply where runoff is likely to occur. Do not contaminate water by cleaning of equipment or disposal of wastes."

These frightening words are quoted from the label of a commonly used soil insecticide manufactured by one of the major chemical companies. Does it sound like something you want to be around—or your wife or children? And to think that thousands of tons of this and equally toxic chemicals are sprayed, dusted, and spread every year all over this country and most other agricultural lands, sends chills up my spine. How about yours?

According to the 1980 *Statistical Abstract of the U.S.,* farms having over $2,500 of sales used chemicals for crop and hay insects on 42 million acres (costing $306 million) and herbicides on 89.9 million acres (costing $356 million) in 1969; while in 1974, the figures were 56.6 million acres treated for crop and hay insects ($596 million) and 114.8 million acres acres treated for weeds ($797 million). That doesn't include millions of dollars spent for chemicals to treat livestock and poultry, nematodes, crop diseases, and for defoliation ($121 million in 1969 and $166 million in 1974). More recent figures for 1978 show that the treated acreage increased to 75.7 million acres for crop and hay insects and 163.5 million acres for weeds, but dollar figures are not given for each separate category (1981 *Statistical Abstract of the U.S.*).

The major use of insecticides on corn is for rootworm control. For years, chlorinated hydrocarbons were used (aldrin, dieldrin, chlordane, heptachlor). Then rootworms that were resistant to these chemicals proliferated. So farmers were told to use increased rates of application of the same poisons, and when that didn't work any more, they were sold other more toxic chemicals. When soybeans were grown on the insecticide-saturated soil, the insecticides were absorbed and showed up in some soybean products. Dangerous levels of dieldrin and its breakdown products were found in poultry. The birds had to be destroyed. The federal Environmental Protection Agency banned the use of aldrin and dieldrin in 1974, and chlordane and heptachlor were banned later.

Questions. Why is it that after 30 years of using pesticides, there are more species of insects classed as pests than before? Why is there DDT in the bodies of every American—in fact, in animals from Antarctica to the North Pole? Why do farmers have worse weed problems now than their grandfathers had? Are toxic chemicals really necessary to control weeds and pests? Why are some animals and plants pests, anyway? What other ways are there to control pests?

Pests. Pests are defined as animals or plants that are harmful to ourselves or our property; so as applied to agriculture, this would include the insect, mite, and nematode pests of crops and livestock, as well as weeds, which are considered to be the worst pests of crops, *and* those diseases of crops and livestock which are caused by microorganisms (viruses, bacteria, and fungi). Nutrient deficiency diseases would not be included, and many experts exclude animal diseases and parasites, although we will include those animal pests.

These pests can cause harm by either feeding directly on crops, stored crops, or animals; by releasing toxic substances; by competing with plants for water, nutrients, or sunshine; by carrying plant and animal diseases; and by weakening a plant or animal and allowing other pests or diseases to get started. The number of pest species is relatively small (less than 1% of the total number of insect species are pests), but the few pest species can sometimes be very common and destructive. In fact, many insects, mites, nematodes, fungi, and bacteria are very beneficial because they eat, parasitize, or cause diseases of the pest species. Even some "weeds" are used for forage in some parts of the world, and if no crops are growing on the land, weeds help prevent erosion and nutrient leaching. In addition, all organisms in nature are part of the food web, or "balance of nature," in which one species uses another for food; so indirectly, everything has some value.

If man had not upset the balance of the food web in so many parts of the world, all or nearly all of the so-called pests would not be pests, but would be held in check by their natural enemies. Most of the worst pests have been carried by man from their original homeland, where their natural enemies controlled them, to some other part of the world, where few if any natural enemies exist. Often their names show this, such as the European corn borer, Japanese beetle, and the Mediterranean fruit fly in the U.S., and the Colorado potato beetle, which has wreaked havoc in Europe.

When is a pest a pest? Is all damage that pests do really important? Should we spray a whole field with poison just because of a few leaves being nibbled? Actually, one-half or more of insecticide spraying is unnecessary. Speaking in terms of immediate dollar loss, if the damage a pest does to a crop or livestock does not exceed a certain level, called either the economic threshold, economic injury level, or critical level of damage, control methods do not pay. Not all pest damage causes serious harm. Most plants can lose up to 30-40% of their leaf area before they are harmed, and hardy plants can outgrow some types of damage. Sometimes leaf damage does not occur until after the grain or fruit is mature, so it doesn't matter. Healthy livestock and poultry can tolerate a certain level of parasite attack. More important, healthy plants and animals can *resist* and ward off diseases and pests. It has been found as far back as 1961 that the basic cause of corn stalk rot is not the attacking fungi, but low sugar production caused by plant stress, including drought and cloudy weather. The stresses "curtail a plant's ability to produce enough sugar to supply the food needed to sustain plant vigor . . . the stalk and roots get shortchanged . . . fungi that normally are warded off invade and destroy the tissues . . . weakening [stalks] and leaving them susceptible to lodging." Balanced soil fertility is one of the recommended ways to reduce stresses (*Successful Farming*, Nov. 1980, p. 36-AD). We will have more to say along these lines later.

Pest control philosophies. Unfortunately, many people's pest control philosophy can be summed up by, "The only good bug is a dead bug!" If you don't believe this, just think of the wild goose chase the pesticide companies have led us on for the last 30 or 35 years. More and stronger insecticides and herbicides. But pest problems have only gotten worse, for the most part, and the ecological side effects have been devastating. *Finally*, now that it is almost too late, some of the "experts" are starting to preach another philosophy.

Pest *control* (also called pest *management*) does not mean pest *elimination*. There is no way every single corn earworm can be killed (more would come in from the neighbor's fields anyway!) or all damage can be eliminated. Thus, in pest control, the main goal is keeping damage below the economic threshold, if possible. There are two ways of doing this: (1) prevention of pest attack in the first place, and (2) control of the numbers of pests, to reduce the intensity of attack. Obviously the first is the most desirable, but both are usually necessary.

There are many ways of attacking pests, some widely used, some largely impractical at present, and some still in the research stage. Using poisonous pesticides is only one, and usually the least desirable in the long run. In modern pest control, often called *integrated pest control,* we attempt to use more than one control method, to attack the pest's weakest points, and to do so in an ecologically sound way. This requires the detailed knowledge of the pest's life cycle and weak points, knowledge of when and how to use different control methods, plus frequent surveys of fields to obtain up-to-date information on local pest abundance.

Control methods. We can divide pest control methods into two large categories, (1) natural control, referring to weather factors and natural enemies of pests (predators, parasites, and diseases); and (2) artificial (applied) control, which refers to any man-directed methods. There are many artificial control methods; some of the more commonly used ones include (a) using man-released natural enemies (such as lady-bird beetles or a bacterial caterpillar disease or parasitic wasps), (b) releasing sterilized males of certain insect pests (such as the screw-worm fly and mosquitoes), (c) using attractant or repellent chemicals or plants (such as sex-attractant odors in an insect trap or spraying garlic oil on your garden plants), (d) planting resistant or repellent varieties of crop plants, (e) cultivation methods that kill weeds and hibernating insect pests, (f) removal of dead plants and spoiled fruit to prevent disease spread and eliminate insect hibernation sites, (g) adjusting planting and harvesting times to escape major pest attack, (h) crop rotation to escape soil-harbored pests and diseases, and (i) using toxic chemicals (poisons) to kill pests.

> There are several terms used to name toxic chemicals used for pest control. The most general would be *pesticide*. Specific types of pesticides would include *herbicides* to control weeds, *fungicides* for fungi, *insecticides* for insects, *acaricides* or *miticides* for mites, *nematicides* for nematodes, and *rodenticides* for rodents and other small mammals.

There is a wide variety of toxic chemicals that are used for pest control, ranging from some that do not cause serious ecological problems to others which are incredibly dangerous poisons related to wartime nerve gases. The least dangerous general group of poisons is often called the *botanicals*; that is, natural chemicals produced by plants. They break down soon (they are biodegradable) and seldom pollute the environment. Commonly used botanicals include pyrethrum (also called pyrethrin), rotenone (this one is toxic to fish), and sabadilla. A second group of toxic pesticides is the *inorganics*, some of which have been used for hundreds of years. There is a wide variety of chemicals in this group; some are relatively harmless, such as sulfur and lime-sulfur, but others are very toxic, such as copper sulfate, arsenic compounds, and mercury compounds. The third group of toxic chemicals is the recently developed (since World War II) *synthetic organics*, most of which are very toxic and do not break down rapidly in the environment (they are called persistent pesticides). Fortunately, many can eventually be broken down by sunlight, oxygen, and bacteria, but some contaminate the entire ecosystem for years. Some have been found all around the world, including in the middle of the oceans, and in the bodies of Eskimos and polar bears. Even human embryos and nursing mother's milk carry pesticide residues. When fish or birds of prey eat food contaminated with pesticides, the poisons can become tens or thousands of times more concentrated in their bodies. Typical amounts of DDT measured in parts per million found in the environment are (from C.A. Edwards, *Persistent Pesticides in the Environment*, 1970, p. v):

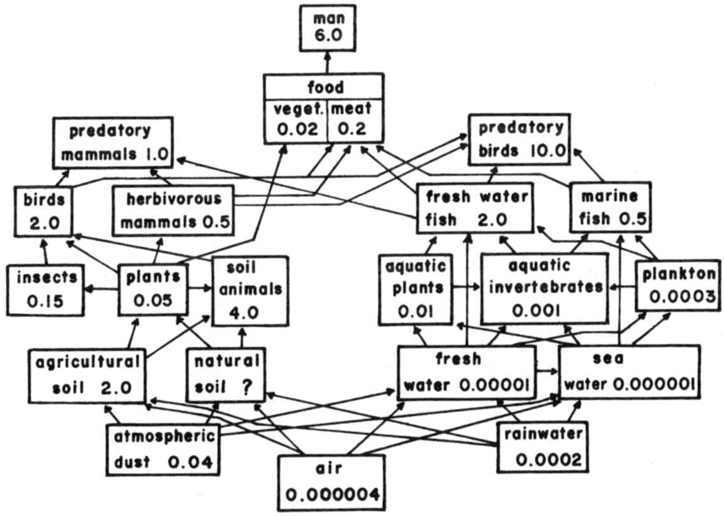

Solutions or problems? The use of toxic herbicides and insecticides has been, and still is (just read the ads in any farm magazine), proclaimed as the savior of agriculture. USDA and state university pamphlets, and even some Secretaries of Agriculture have painted a picture of starvation and doom if chemical pest control is not followed, using such figures as . . . without chemicals crop production would fall 30% and food prices would rise 50-70%. But what have been the results of over 30 years of heavy use of toxic pesticides? Consider these facts:

1. DDT, DDE, and dieldrin were found in the fat tissue of 100% of 54 cadavers tested by researchers in Dade Co., Florida. Those who were heavy users of household pesticides had over four times the DDT and nearly three times the DDE than those who were low users. The levels of DDE and dieldrin were much higher in those people who had died of brain disease and cirrhosis of the liver (*Chemical Fallout*, 1969, p. 301, 308-309).

2. After the herbicide 2, 4, 5-T (Agent Orange) was used for defoliation during the Vietnam war, birth defects increased greatly in the native population (H. Wellford, *Sowing the Wind*, 1972, p. 195-196). In 1970, drift from U.S. Forest Service spray in Arizona caused cases of stillborn children and deformed goats and chickens. Tests run for the National Cancer Institute found that rats produced 39% abnormal fetuses at a low dose and 90-100% at "relatively high doses" (R. Harmer, *Unfit for Human Consumption*, 1971, p. 93). A contaminant of 2, 4, 5-T is the chemical dioxin. In 1976, an explosion at the Icmesa chemical plant sent a cloud of dioxin over the northern Italian town of Seveso. About 700 people near the plant were evacuated. Within weeks, hundreds of villagers developed a skin disorder, chloracne. In 1979, the number of women who had been exposed to dioxin that died of breast cancer was double the national average. The number of babies born with deformities rose from 3 in 1975 to 53 in 1978. Farming was banned on 4500 acres. In 1982, the Italian government lifted the ban on farming in some areas (soil tests showed no "toxicity"), but in order to reopen the road from Seveso to Meda, the soil along the road must be turned over. The government won't release details on soil and medical tests (*Newsweek*, May 10, 1982, p. 14, 18).

3. In 1961, an organic phosphate insecticide, phosdrin, accidentally was spilled on a bundle of bluejeans being trucked from Fresno to Los Angeles, California, in the same truck. Six children who wore the jeans were stricken with such symptoms as rapid heartbeat, irregular breathing, glassy eyes, muscle twitching, diarrhea, nausea, vomiting, and abdominal pain (R. Harmer, *Unfit for Human Consumption*, 1971, p. 21-23). Few farmers or homeowners read the label before using pesticides. Safety standards are commonly violated. Farm workers and fruit pickers are often sprayed by planes (H. Wellford, *Sowing the Wind*, 1972, p. 229, 232). Sixteen out of 24 vineyard workers were overcome when they entered a vineyard in California 33 days after it had been sprayed with an

organic phosphate (nerve gas type) insecticide (the permissable time limit in California was 14-28 days). Vegetable farmers anxious to deliver flawless produce to market have been known to douse their crops with incredible amounts of insecticide; one lettuce grower was caught using chlordane, endrin, dieldrin, DDT, toxaphene, malathion, cryolite, and rotenone only 11 days before harvest (B. Mooney, *The Hidden Assassins*, 1966, p. 106). The U.S. Food and Drug Administration's budget and staff are so small that less than 1% of interstate food shipments can be inspected (R. Harmer, *Unfit for Human Consumption*, 1971, p. 40, 94).

4. U.S. export trade is sometimes harmed because many foreign countries no longer accept pesticide-contaminated U.S. food and tobacco. The European Common Market countries set much lower pesticide residue tolerance limits than the U.S. In 1969, the tolerance limits of several insecticides were:

	European (parts per million)	*U.S. (parts per million)*
Captan	15.0	100.0
Lindane	2.0	10.0
Parathion	0.5	1.0
Toxaphene	0.4	7.0

(R. Harmer, *Unfit for Human Consumption*, 1971, p. 37, 40).

5. *Combinations* of toxic chemicals are often more dangerous than single ones (this phenomenon is called synergism). Combinations of chlorinated hydrocarbon insecticides at "harmless" levels caused liver damage in test animals, but did not when administered singly at the same dose (*Quarterly Bulletin of the Association of Food and Drug Officials of the U.S.*, Fall 1950, p. 90). The insecticide malathion, considered a "relatively harmless" organic phosphate, became over eight times more toxic when administered to animals along with EPN (*Journal of Pharmacology*, vol. 121, p. 96).

6. Over 200 species of insects and mites have developed a resistance to pesticides (*General and Applied Entomology*, 3rd edition, 1972, p. 454)—a more recent figure is 400 (*U.S. News & World Report*, June 14, 1982, p. 44). "Superflies" can tolerate a dose equivalent to feeding 14 pounds of poison to a 200 pound man (R. Harmer, *Unfit for Human Consumption*, 1971, p. 58). The German cockroach is so resistant to both chlorinated hydrocarbons and organic phosphates that it has been called "indestructible" (*Wall St. Journal*, Feb. 14, 1968).

When corn prop roots are reluctant to enter the soil, herbicide carryover is to be suspected.

7. Pesticides kill many organisms in the ecosystem, not just the "target" pests. Aldrin and dieldrin greatly decrease populations of soil arthropods, while endrin kills many earthworms (*Residue Reviews*, 1975, p. 40-41). Pesticides can affect the metabolism of plants, and plants can accumulate pesticides from the soil *(Residue Reviews*, 1974, p. 73-74). Over 30 currently used pesticide preparations are labeled as toxic or higher toxic to honeybees (Iowa State University Entomology Extension sheet). Some pesticides are very stable and persist in the environment for many years (others break down rapidly); for example, after 14 years the following percents of pesticides were still present in a sandy soil: aldrin 40%, chlordane 40%, endrin 41%, heptachlor 16%, dilan 23%, isodrin 15%, BHC 10%, toxaphene 45% and (after 17 years) DDT 39% (*Residue Reviews*, 1974, p. 109-110). Persistent pesticides are passed along in the food chain and *concentrated* as they go: in Clear Lake, California, small amounts of DDD (14 parts per billion) were used for gnat control. The microscopic plankton organisms were found to have 5.3 parts per million (ppm) DDD, plant-eating fish had 40-300 ppm, carnivorous fish had 2,500 ppm, and birds had 1600-2000 ppm (R. Harmer, *Unfit for Human Consumption*, 1971, p. 62).

This list of "horror stories" could go on and on, but I think you get the point: *the use of pesticides is not ecologically sound*. William Longgood sums it up well in his book, *The Poisons in Your Food* (1960, p. 255): "The fact that sprays are claimed to be necessary to grow food is not a warrant for their increased use; it is an indictment of the critical plight of American agriculture and the waning fertility of our once-rich soils. Poison sprays, rather than being a solution, are merely a delaying action, a final desperate maneuver to stave off catastrophe."

There HAS to be a better way.

Nature's way—Weeds. By working *with* natural systems, effective control of pests and weeds can be obtained. The basic *cause* of pests is out-of-balance soil, believe it or not. Let's first consider weeds. Weeds are undesirable plants (carryover corn in a soybean field could be called a weed). Most weeds are "opportunistic" plants, growing best in poor soil and waste places. Some species do best in very acid or alkaline soil. Weeds compete with crop plants for water, light, and soil nutrients.

Studies have correlated weeds with out-of-balance soil: sheep sorrel indicates a need for lime, foxtail barley indicates saline soil, and dandelion and plantain grow well in alkaline soil (alkaline-producing fertilizers such as sodium nitrate promote their growth) A. S. Crafts & W. W. Robbins, *Weed Control*, 3rd edition, 1962, p. 89). Experiments by Dr. William Albrecht and his associates at the University of Missouri showed that fields fertilized with barnyard manure were immune to

weeds. Dr. Ehrenfried Pfeiffer has correlated weeds of the mustard family with improper fertilization, especially with excessive salt forms of potassium (C. Walters, Jr. & C. J. Fenzau, *An Acres U.S.A. Primer*, 1979, p. 295-297). Anaerobic soils and wrong decomposition of organic matter, along with death of beneficial soil microbes, stimulates certain weeds, including buttonweed, jimson weed, stinging nettle, Johnson grass, quackgrass, and bindweed (certain soil fungi digest bindweed and morning glory rhizomes and roots) (*An Acres U.S.A. Primer*, p. 300-301, 317, 322, 345).

Some weeds are encouraged by the use of herbicides and wrong fertilizer materials (which can throw the soil out of balance and kill beneficial organisms), including foxtail (helped by the high magnesium of dolomitic limestone, which tightens the soil) and fall panicum (allowed to grow when herbicide treatment removes foxtail) (*An Acres U.S.A. Primer*, p. 181, 306, 308). One of the most common causes of weeds is high potassium soil.

Most weeds actually do not grow well in good, healthy soil and in competition with vigorous, healthy crop plants (*Weed Control*, 3rd edition, p. 93). I have seen sick, insect-riddled weeds with hollow stems and rotting roots growing next to healthy corn and alfalfa in fields that have been under a corrective fertilization program for only 7 months. A few weeds will grow in good soil, but are seldom a serious problem; two weeds that indicate good soil are lambsquarters and redroot (rough) pig-

No herbicide was used on this corn field.

weed (*An Acres U.S.A. Primer*, p. 322-324); in some parts of the world these "weeds" are a valuable forage crop. Certain plants actually exude toxic substances from their leaves and roots, usually while in the seedling stage, which inhibit the growth of, or even kill, neighboring plants. This phenomenon is called allelopathy. For example, quackgrass has been found to inhibit alfalfa seedlings, and oats inhibit growth of the weed charlock up to 38% (*Weed Control*, 3rd edition, p. 103-107). There is some interest among researchers in using allelopathic plants or their toxins as natural herbicides (*Crops & Soils Magazine*, March & April 1980).

Non-toxic weed control. Weeds don't have to be a problem. Effective and ecologically sound weed control methods can do the job. First and most important, build up your soil to a proper balance of nutrients, tilth, and aerobic organisms. This will provide optimum conditions for crops and poor conditions for most weeds. This plus preparing a good seedbed in the rows and not between, will allow the crops to sprout first, get a head start on weeds, allow allelopathy to operate, and allow crops to shade out weeds. Later, remaining weeds can generally be controlled by cultivation. Cultivation is such an effective means of weed control that in 1941, the perennial weeds on about 40,000 acres in Idaho were nearly or completely eradicated (*Weed Control*, 3rd edition, p. 148), yet do we hear much about it in these days of "better living through chemistry"?

An alternate strategy that often works well is to delay planting until the weeds have germinated and then kill them by cultivation while the seedbed is being prepared. I have seen this method produce essentially weed-free cornfields; the neighbors asked what herbicide was used. In smaller areas, frequent mowing can control weeds because they will not be able to make enough food to sustain themselves, and the roots will eventually die. Also, mulching will provide effective weed control for gardens and truck crops.

Nature's way—"bugs". Just as the ultimate cause of weed problems is out-of-balance soil, so is it for insect, mite, and nematode pest problems, and plant diseases, for in general these pests attack unhealthy, sick plants and leave the vigorous, healthy ones alone; and of course, the main cause of sick plants is "sick" soil. It may seem hard to believe that insects and diseases attack unhealthy plants and leave the healthy ones alone, for some farmers' crops and gardens never seem to escape attack, but there are many farmers and gardeners who follow ecologically sound methods and can vouch that it is true; and scientifically performed experiments also prove it. Having had a traditional university education in entomology, I also could not believe it until I saw it.

Accounts of armyworms detouring around one farmer's field of oats and destroying his neighbor's, or of one field of corn being attacked by corn borers but not another, are not just "fish stories"—they are true. I have seen one field of oats in healthy condition, while all those around it were being decimated by aphids and red leaf disease in 1982. What could make the difference between two neighboring fields, one decimated by insects or disease and the other untouched? Of course it is true that there are some *varieties* of crop plants that are more resistant to pests than others. The classic book *Insect Resistance in Crop Plants* by R. H. Painter (1951), gives many examples. But it also mentions other factors that contribute to pest resistance, including the plant's production of substances that are repellent or toxic to insects, *and* general vigor and soil fertility. The presence of non-lethal plant diseases (in other words, poor health) increases insect attack. A study by P. Schaper found that Colorado potato beetles prefer to lay eggs on leaves of plants that are in poor condition and that plants with virus infections furnish better food for the insect. Dr. Painter surveyed experiments relating fertilizer to insect attack but could find no pattern; some tests found that an increase in a nutrient (say, nitrogen or phosphorus) decreased insect attack, while others found that it increased attack. Therefore he doubted that soil fertility

Plant "Antibodies"

An amazing method that plants use to combat diseases has recently been discovered. If an invading fungus, bacteria, or nematode succeeds in killing a few cells, the *surrounding* cells produce a substance called a phytoalexin, which is toxic to that particular disease pathogen and usually not others, similar to the way animals' bodies produce antibodies to fight a specific disease. Dr. T. J. Wacek, writing about phytoalexins in the January 1977 issue of *Organic Gardening* (p. 132-134), expresses hope that they can be developed for commercial use, since they do not kill helpful microorganisms: "Fungicides are the worst offenders in disrupting balanced colonies of beneficial microbes in living soil." Three Canadian researchers have tested a phytoalexin from peppers and found that it effectively controls a fungus causing late blight in tomatoes (*Phytopathology*, vol. 65, Feb. 1975, p. 68-69).

plays an important role. The answer probably is that it is the *balance* of many nutrients, not just the amount of one, that is important, plus the general "health" of the soil (humus, tilth, microorganisms). Too much of a nutrient can upset a plant's metabolism just as easily as a deficiency. For example, *excess* nitrogen decreases the tolerance of sorghum to chinch bugs, while adequate phosphorus increases tolerance (R. G. Dahms, *Journal of Economic Entomology*, vol. 33, no. 4, p. 688-692). Dr. William Albrecht and his associates found that adequate amounts of both nitrogen and calcium made spinach plants immune to thrips, while plants fertilized with low levels were attacked. (S. H. Wittwer & L. Haseman, *Journal of Economic Entomology*, vol. 38, no. 5, p. 615-617). Nitrogen is necessary for protein production and calcium aids in nitrogen utilization. Albrecht assumed that the well fertilized plants were able to produce certain proteins that act as antibiotics against insects (*The Albrecht Papers*, 1975, p. 260-261).

Some researchers believe that some amino acids which are not part of proteins protect plants from disease pathogens (J. A. Goss, *Physiology of Plants and their Cells*, 1973, p. 201-204). One study found that wheat resists rust when grown at cooler temperatures (64°F daytime, 35° F night) AND that the resistant wheat contained at least two proteins that susceptible wheat did not have (*Phytopathology*, 1965, p. 413-414). Another study markedly increased the resistance of 7 apple varieties to apple scab by treating plants with the amino acid phenylalanine (*Phytopathology*, 1959, p. 313-315). Prof. Albert Schatz of Washington University, has said, "Whether insects prefer or do not prefer to feed on certain plants may be determined by the kinds and amounts of amino acids in those plants. And the kinds and amounts of amino acids present may, in turn, be determined by the mineral nutrients available to the growing plants. Thus soil fertility can influence the resistance or susceptibility of plants to attack by insects" (*Organic Gardening and Farming*, Feb. 1969, p. 22). It has been found that chlorine reduces amino acids and protein snythesis, and promotes protein decomposition (*Annual Review of Phytopathology*, vol 12, 1974, p. 151). Dr. Albrecht also found that when calcium and potassium are in proper balance, potatoes are able to use copper to produce antibiotics that protect against the potato scab fungus, and he cites experiments in South Dakota proving that corn was protected from the lesser grain borer when nitrogen and phosphorus were in balance (*The Albrecht Papers*, p. 201-202, 359).

Either an excess or a deficiency of soil moisture can cause plants to be more susceptible to insect attack (*Insect Resistance in Crop Plants*, p. 91). Walters and Fenzau report that high magnesium (remember dolomite?) favors wireworm attack because of its adverse effect on soil moisture tension (*An Acres U.S.A. Primer*, p. 276). This, plus the effects of unbalanced nutrition indicate that such environmental and functional *stresses*—in other words, poor health—are a major cause of plants being attacked by pests and diseases.

Another part of the story is the help given to plants by some of the beneficial soil microorganisms. Certain bacteria and fungi which live near plant roots produce powerful antibiotics that protect plants from parasitic fungi (E. W. Russell, *Soil Conditions and Plant Growth*, 9th edition, 1961, p. 228-232; S.D. Garrett, *Biology of Root-inhabiting Fungi*, 1956, p. 116-124). (See box, Chapter 3.) The special symbiotic fungi, mycorrhizae, also can produce antibiotics to protect roots from disease organisms (A. J. R. Guttay, *Crops & Soils Magazine*, Aug.-Sept. 1975, p. 15). Several species of certain types of soil fungus trap and devour nematode worms (see box).

How bugs "tune in" to sick plants. The way in which insects locate the plants they intend to eat or lay eggs on (so their offspring can eat) is extremely fascinating. The whole story is just now being discovered. Most pests prefer to eat only certain types of plants, sometimes only one species; for example, the European corn borer prefers corn but will also attack sorghum and beets, while the onion maggot only attacks onions. But how does a female corn rootworm beetle tell a corn plant from a soybean at a distance of 100 feet, and even more amazing, how does she tell a sick corn plant from a healthy one? Entomologists used to assume that insects located objects by odor, vision, taste, and feel.

But a USDA researcher, Dr. Philip Callahan, is opening up a whole new dimension in insect activity. Dr. Callahan, who has a background in electronics and radio, has found that some of the tiny hairs on the antennae of insects are sensitive to varying wavelengths of infrared. Now, everything in nature radiates ("broadcasts") invisible infrared wavelengths depending on its temperature and molecular composition. Odor molecules broadcast at different wavelengths depending on their type, concentration, temperature, and vibrations. Thus insects can easily distinguish between different kinds of plants and other objects in their environment. Dr. Callahan describes these things in his book, *Tuning in to Nature* (1975). It is known that sick plants have a different temperature than healthy ones (*Successful Farming*, Dec. 1981, p. 26-P), and it is reasonable to assume that a sick plant with upset metabolism will produce different odor molecules than a healthy one. Thus sick plants will "look" different at infrared wavelengths to an insect. Man already uses

Predatory Fungi

A science fiction writer could hardly come up with a stranger tale than the story of soil fungi that "prey" upon soil animals, especially the nematodes (roundworms, threadworms), which are often serious plant parasites (not all nematodes are harmful pests; many are predators in the soil or else live on decaying organic matter; others however are animal parasites).

Some of these fungi produce tiny donut-like "nooses" at intervals along their filaments. When microscopic animals, including potentially harmful nematodes, poke their bodies through, the "noose" instantly constricts, trapping them; then the fungus dines on its hapless victim. A nematode seldom escapes. Other species of fungi trap nematodes on sticky knobs and branches (C. L. Duddington, *The Friendly Fungi*, 1957; G. L. Barron, *The Nematode-destroying Fungi*, 1977). One type grows nematode traps only when nematodes are in the vicinity; otherwise it lives without "fresh meat" (W.A. Jensen & F. B. Salisbury, *Botany: An Ecological Approach*, 1972, p. 365).

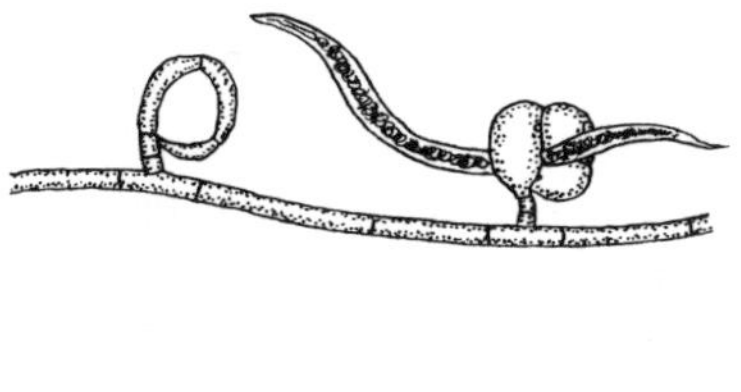

Unconstricted and constricted ring traps.

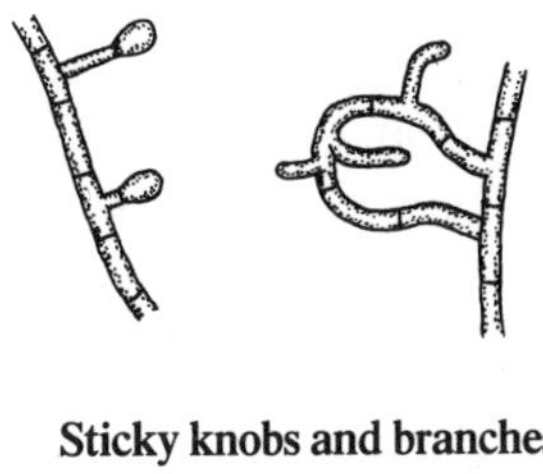

Sticky knobs and branches.

this to survey crops by taking infrared photographs from airplanes and satellites. We are just duplicating what the insects already do. Dr. Callahan is currently doing research to definitely prove that insects *prefer* sick plants over healthy ones.

Two mechanisms could operate, or both at once. Either sick plants *attract* pests, or healthy plants *repel* them, or both. It is known that aphids are attracted to unhealthy plants which have a high sugar content. The reason for the high sugar is that the plant's metabolism and "plumbing" are unable to convert sugar to other materials and unable to translocate it from the leaves where it is made. On the other hand, a healthy corn plant produces a substance called dimboa, which repels the European corn borer. The corn borer is usually a problem on soils with a low humus level, too much magnesium, poor tilth, and a low pH (*An Acres U.S.A. Primer*, p. 275, 286).

Non-toxic "bug" control. So, to make a long story short, control of insects, nematodes, and plant diseases is as simple as weed control—good, healthy, fertile soil in order to reduce stresses on crops, provide beneficial microorganisms, and produce healthy, vigorous plants. You shouldn't have to do anything else in most years. But sometimes other control measures may be necessary, say because your soil isn't yet in "good health" or because unusual weather conditions put added stress on your crops.

If pests do begin to attack your crops, don't immediately rush to your local "poison dealer." First of all, the damage may not be above the economic threshold; it may not be serious enough to worry about. Wait and watch. Frequently the weather or natural enemies of the pest (predators, parasites, diseases) will eventually take care of the problem (it may take a couple of weeks, though). I have seen tiny parasitic wasps eliminate cabbageworms on a few unhealthy cabbage plants in an otherwise healthy garden. The wasps, resembling gnats, swarmed around the cabbages. The females laid eggs in the cabbageworm caterpillars. The wasp larvae ate the blood and internal organs of the cabbageworms, finally emerging to spin white cocoons on the leaves and change into adult wasps.

If the pests seem to be getting out of hand and are doing serious damage, other control measures may be effective. Non-toxic methods should be considered first. For some pests, you can release natural enemies (such as parasitic wasps, ladybugs, or a bacterial disease of caterpillars); again, this may take a couple weeks to work. Sometimes light or odor traps, or barriers will reduce numbers or keep pests away. Sprays of repellent odors or dusts of gritty materials (wood ashes, diatomaceous earth) are often effective. Gritty materials scratch insects' cuticles and

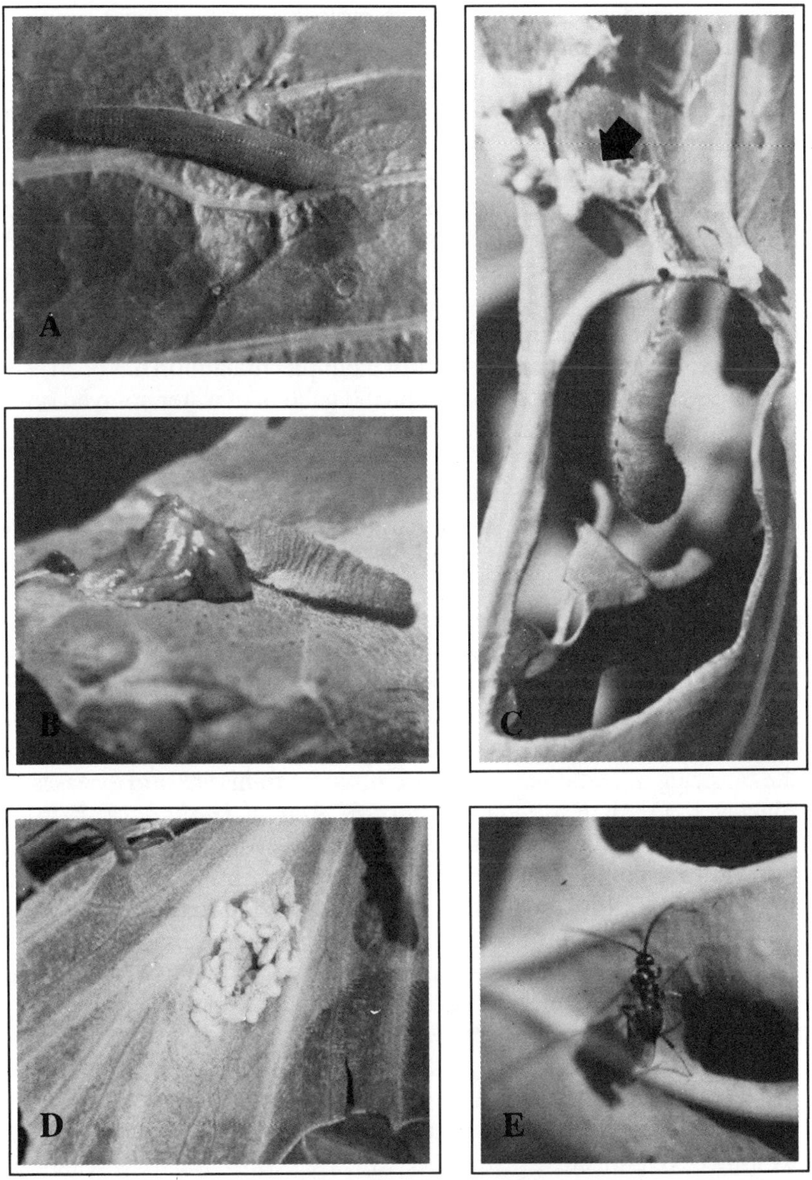

***Macrocentrus*, a tiny ($^1/_8$ inch) wasp parasite of the cabbageworm (European cabbage butterfly). A. normal-looking cabbageworm; B. cabbageworm with many wasp larvae squeezed out; C. dying cabbageworm with wasp cocoons on leaf (arrow); D. cluster of wasp cocoons; E. adult wasp searching for new cabbageworms in which to lay eggs.**

they die from dehydration. Spraying molasses or other sugar solutions will make plants unpalatable or will suffocate small pests.

Toxic chemicals should be used only as a last resort. If possible, use the less dangerous natural plant-derived substances (botanical insecticides), such as pyrethrum, rotenone, and sabadilla. The less toxic inorganics can also be relatively safe (sulfur, lime-sulfur) and may even help fertilize your plants. The more toxic and environmentally dangerous inorganic and synthetic organic pesticides should be avoided. Sometimes, however, the "relatively safe" toxic pesticides can be used effectively at much lower rates of application than what the label recommends (apparently the recommended doses are just below what will kill the plants and far above what will kill the pests). For example, malathion has been used on snapbeans at a rate of 1 ounce in 100 gallons of water sprayed on 20 acres, while the dosage recommended on the label is 1/2 ounce per gallon.

Animals too. Just as good health and vigor protects plants from their pests and diseases, so also are animals (and humans) protected from parasites and diseases. Are infectious diseases *caused* by "germs"? Well, yes *but* . . . we are surrounded by disease germs daily, but as long as we are in good health—get plenty of sleep and eat a good diet—we don't get sick. Usually, it is only when an animal is under stress—in poor health—that disease pathogens can get a foothold. Healthy animals have various defenses against parasites and diseases, including antibodies and white blood cells. It is well known among animal breeders and geneticists that the offspring of certain crossings are resistant to insects and diseases (J. Blakley & D. H. Bade, *The Science of Animal Husbandry*, 1976, p. 129). Organic farmers often report that their livestock are not bothered by flies. Veterinarian Dr. John Whittaker, writing in *Acres U.S.A.* (Dec. 1975, March & April 1976), states that B vitamins, vitamin C, and other nutritional factors play an important role in protecting animals from parasites and diseases; for example, an imbalance of dietary calcium and phosphorus or a magnesium deficiency increase parasitic worm infestations, while a high carbohydrate diet increases the infection of *Balantidium*, a protozoan intestinal parasite. He notes that too much soluble nitrogen (non-protein nitrogen) or urea in feed causes high blood urea or ammonia levels, leading to reduced resistance to bacterial infections. Resistance to parasites and diseases can also be lowered by vaccinations and antibiotics (these can kill rumen microbes, leading to toxic mold infections), worming medicines, moldy feeds (through mold-produced toxins, including aflatoxin), and stresses (weather, noise, moving, and diet changes).

Which for you? The basic approach of the experts to weeds, insects, and diseases is to identify the pest involved and "zap" it with the recommended poison. But actually, pest attack is a symptom of plant and animal deficiencies and malnutrition, not the cause of illnesses.

We have seen that it is possible to do without the chemical arsenal of insecticides, herbicides, fungicides, and all the other -ides (they're basically SUICIDE!). If natural systems are simply *allowed* to become reestablished and function as they were designed, any pest, weed, and disease problems will be taken care of, and more important, if the soil is in good, healthy condition, most pest problems won't even occur, they will have been *prevented*.

Weeds, insect "pests," and diseases actually do us a valuable service. They were put on the earth to eliminate the sick and unfit, so they tell us that something is wrong—with our animals' health, with our crops, and ultimately with our soil. Will we heed their warning?

CHAPTER 17

Quality

Which would *you* rather eat, a nice plump, red, juicy, *sweet* apple or a shriveled, green, bitter, half-rotten thing called an apple by your grocer? The answer is obvious; you want quality food for yourself. So why do most farmers feed miserably poor quality, vitamin and mineral deficient, and even toxic feed to their animals? And then they have to add supplements to fortify it and molasses to get the critters to even eat it!

The general trend of western civilization seems to be away from quality and toward "just enough to get by." This is clearly indicated by the flood of cheap plastic throwaway products on the market. Even in agriculture, the emphasis for so many years has been on quantity—bushels and pounds—and not quality.

Dr. William Albrecht wrote that in 1911, the protein content of corn was 10.3% (this was before hybrids). In 1950, five hybrid varieties were tested; the protein content ranged from 7.9 to 8.8%. In 1956, 50 hybrid varieties averaged 5.15%. The same trend has occurred in other crops; in 1940, Kansas wheat averaged nearly 19% protein; it dropped to 14% in 1951 and 10.5% in 1969. These are shocking losses in quality. At the same time, yield per acre has steadily risen. Some average yield figures from the 1969 *Statistical Abstract of the U.S.* and the 1981 USDA *Agricultural Statistics* are:

Corn, bushels/acre		*Wheat, bushels/acre*	
1931–35	22.8	1931–35	13.1
1941–45	32.8	1941–45	17.5
1951–55	39.1	1951–55	17.9
1957	47.1	1957	21.7
1960	54.7	1960	26.1
1966	73.1	1966	26.3
1969	85.9	1969	30.6
1973	91.3	1973	31.6
1979	109.7	1979	34.2

Such are the marvels of modern agriculture: more bushels per acre, lower nutritional value, sick animals, higher veterinary bills!

What do we mean by quality? How is quality measured? Is high quality really important? Why take the trouble to strive for quality?

Quality—what is it? The dictionary defines quality as degree of excellence; superiority; capability of producing specific effects. That can certainly be applied to agriculture, but it depends on what criteria are chosen to measure quality. Some are very misleading. For example, animal nutritionists and feeds experts usually measure crude protein (CP). This may sound like an overall measure of protein content, but in practice, the lab technician only measures the % of nitrogen in the ashes of the sample and multiplies by 6.25 to give "crude protein." This method is based on the *assumption* that the protein content (which is harder to measure) is 6.25 times greater than the nitrogen content. This may or may not be true (Research Report 29, Part 1, Univ. of Wisconsin College of Agriculture Experiment Station, Oct. 1967, p. 2). If the metabolism of the plant was upset, protein synthesis may have been slowed or blocked, and substances preliminary to protein formation may have built up in the plant's tissues: nitrate, ammonium, amides, and amino acids. These would give a higher nitrogen reading, but they are not protein. And protein content isn't the whole story. The proper *balance* of amino acids in the proteins is necessary for good health in animals.

We have covered some methods of measuring nutritional value in Chapter 5. There are many methods, some easy, some hard, some better, some misleading. Yield in bushels or pounds per acre is of course not a reliable measure at all. That is why we are growing bumper crops of "foodless food." A somewhat better measure is test weight. For decades the standard weight of a bushel of corn was 56 pounds. Then so much poor quality corn was being grown that the weight of number 2 corn was lowered—and lowered—to 53 pounds last year—and 52 pounds this year. No, bushels haven't gotten smaller, corn is getting lighter. Maybe we're growing cornflakes! Yet *some* farmers are growing 60 or 61 pound corn. They have their soil in balance.

The best measure of food quality is what it does for the animal (or person) that eats it. Does it cause good weight gain, milk or egg or meat production, normal reproduction, vigor and disease resistance? Decades ago, the standard amount of feed needed daily by an average dairy cow was 32 pounds. Today, 50 pounds are considered necessary. Maybe more is needed for greater milk production, but why are protein and mineral supplements "necessary" now? Are we growing more bulk—"belly filler"—and not nutritious feed? Farmers with in-balance soil report that their cattle eat only one-half or two-thirds as much feed, but give more milk with higher test figures, and the incidence of mastitis and other ailments drops or disappears completely. (For example, a northeastern Wisconsin dairyman, after beginning to feed his herd feed grown on balanced soil, was pleased to find his average butterfat figure rise from 540 pounds in May of 1982 to 555 pounds in July, and his rolling herd average go from 15,260 pounds in May to 15,499 pounds in July.)

A quick, easy way to get an indication of crop quality while the plants are still growing is to use an instrument called a *refractometer*. A refractometer is a relatively inexpensive hand-held device which measures the sugar content of a couple drops of any liquid. Refractometers are commonly used in industry as a quality check: by sugar beet buyers to determine grade and price paid to growers, by wine makers to check grape quality, and by alcohol distillers to test grain. They are used by some doctors and veterinarians to test urine. You simply squeeze a few drops of juice from the stem or leaves of your crop plant onto a glass

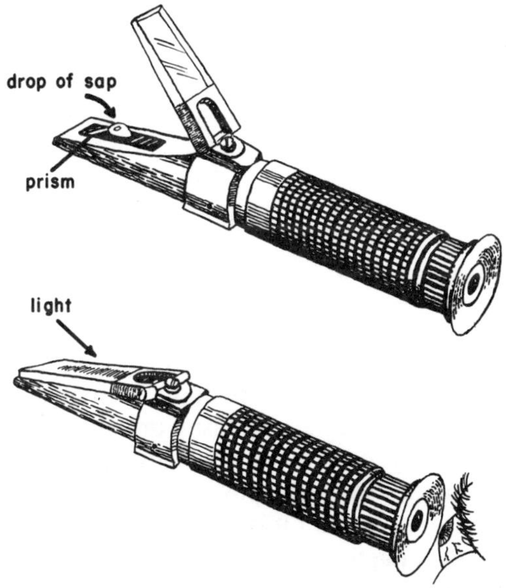

Using a refractometer is easy. A drop or two of sap from the plant is squeezed onto the glass prism. Then the "lid" is closed; the instrument is held in the light, and the sugar content is read on a scale visible through the eyepiece.

prism in the refractometer and then look through the eyepiece. The sugar content is read on a numbered scale in units called Brix (same as per cent). By comparing with standard levels (see box) and with past readings for your crops, you can see how well your crops are functioning—making sugar. You should be careful to test juice from the same part of a plant each time, say the corn stem at the third leaf, or at the cob level, since sugar content does vary along the length of the stalk. Also, readings will be higher on sunny days. Although the refractometer only measures sugar content, it is still a good (and easy) way to measure quality and feed value, since tests have shown that protein and mineral content are directly correlated to Brix readings. You can even use the refractometer to test the quality of fruit and vegetables you buy at the store (although the grocer may take a dim view of it).

Brix (sugar) Readings for Some Plants

	poor	*average*	*good*	*excellent*
alfalfa	4	8	16	22
carrots	4	6	12	18
grains	6	10	14	18
corn stalks	4	8	14	20
corn, young plant	6	10	18	24
grapefruit	6	10	14	18
lettuce	4	6	8	10
sweet corn	6	10	18	24
watermelon	8	12	14	16

Still another good indication of quality for many crops is their storage and keeping ability. People who store fruit and vegetables certainly know that in some years, produce keeps better than others, or that crops from certain fields store better. If grain isn't dry enough, it will mold in storage. Generally today, corn quality is so low that farmers haul their corn to the local mill or elevator and pay to have it dried by using expensive fossil fuel. How much easier it is to simply "grow" your grain dry; that is, to have healthy plants which finish the normal physiological development of the grain, so that the water content of the maturing grain is naturally reduced at the same time that the kernel is filling up with food and minerals. Such completely matured grain will store mold-free at a higher moisture content than incompletely matured grain (what moisture there is in mature grain is "bound moisture," locked into the stored food molecules; it will not cause mold problems). In order for this to occur, the soil and plant have to be in top condition at the end of the

growing season, and the "plumbing" of the plant must not be plugged up or degenerated (see Chapter 4). Believe it or not, corn doesn't have to dent; even "dent" varieties will not dent as much as usual when they are of high quality and full of minerals and protein. Moisture content and storage quality are correlated with the sugar content, easily measured with a refractometer.

But besides all these ways for man to measure quality of crops, probably the best overall "instrument" to let you know the quality of your crops is your animals. As Dr. William Albrecht has put it, "Cows are capable chemists." Animals have much better senses of smell and taste than we do, and they have the instinct to "know" what nutrients they need. Every farmer who feeds livestock or poultry has probably noticed that animals prefer some feeds over others, or hay from one field. A study in Australia found that sheep ate more clover hay from test plots with higher phosphorus contents (*Australian Journal of Agricultural Research*, 1971, p. 941–950), and another study found that sheep preferred tall fescue hay fertilized with phosphate and moderate (50 pounds per acre) levels of nitrogen (*Journal of Animal Science*, 1965, p. 615–625).

Aflatoxin

Aflatoxins are several highly toxic substances produced by two common molds, *Aspergillus flavus* and *Aspergillus parasiticus*, when they grow on certain foods and feeds. The name comes from *Aspergillus flavus*-toxin. One type of aflatoxin, B_1, is the most powerful cancer-causing agent known. Even at the lowest detectable amount, 0.1 part per billion, scientists are concerned about its hazards in human and animal foods. Other less dangerous toxins are produced by certain other fungi.

Aflatoxins are produced when the two *Aspergillus* species grow under warm, moist conditions on various crops: peanuts, corn, wheat, oats, barley, rice, sorghum, tree nuts (pecans, Brazil nuts, pistachio nuts), coffee, black pepper, and feed meals (cottonseed meal, soybean meal, linseed meal, and fish meal). These foods can be contaminated while in storage or crops can become moldy in the field. Aflatoxins are mainly found in tropical and subtropical areas, and in the U.S. are most common in the Southeast. Abnormal or unhealthy crops are most often attacked: cracked kernels of stored high moisture grain or crops damaged by hail or early freeze, or suffering stress from drought and insect damage. In 1977, there was a serious outbreak of aflatoxin fungi in the Southeast on corn stressed by drought and insects; as much as two-thirds of the crop in some states was contaminated (Joseph Rodricks, "Aflatoxins," *FDA Consumer*, May 1978, p. 16-19).

The danger of aflatoxin to animals and humans was not recognized until 1960. In larger doses animal poisoning and other side effects result. Symptoms include impaired liver functions, reduced blood clotting, fragile capillaries and hemorrhaging, kidney damage, impaired immunity to diseases ("lazy" white blood cells which do not destroy bacteria as they should), interference with vitamins A and D and with calcium metabolism, anemia, reduced weight gain and production of milk or eggs, susceptibility to diseases (pneumonia, mastitis, foot rot, emphysema), and poor reproduction (low conception rate, false pregnancies, abortions, and still births). A long low-level exposure results in liver cancer in more susceptible animals.

continued . . .

Dr. Albrecht points out that cattle avoid eating the taller and greener grass growing around manure "pies" because in that small area, the plants' fertility is out-of-balance, with *too much* nitrogen being provided by the manure. The lush grass which looks better to us is actually of lower nutritional quality to the animal. Dr. C. D. Story of the Colorado State University Agricultural Experiment Station, found that "animals fed nitrogen-fertilized hay do not make as high gains as those fed unfertilized hay" because "fertilizers that provide quickly soluble nitrogen

Aflatoxin, continued

In 1960, thousands of young turkeys, pheasants, and ducks in England died; the cause was traced to a shipment of Brazilian peanut meal used as a protein feed supplement. In 1974, in India, 400 people were poisoned by eating contaminated corn; 106 died. Unusually high liver cancer rates in Africa, India, and the Philippines are thought to be connected with aflatoxins ("Aflatoxin and Other Mycotoxins: An Agricultural Perspective," Report No. 80, Council for Agricultural Science & Technology, 1979).

Aflatoxins cannot be detected by ordinary means. Not all moldy grain or feed contains them. A quick screening check is often run by examination under ultraviolet light; aflatoxin-contaminated materials will glow greenish-yellow, but a further chemical analysis is required for a positive identification.

When animals consume aflatoxin-contaminated feed, most of the toxin is excreted, but some will appear in milk or eggs, and some will be stored in the liver and muscle. It can also find its way into cheese and other dairy products. Sweet corn (fresh, frozen, or canned) is not susceptible to contamination, nor are corn starch or refined corn or peanut oils, but other corn and peanut products are suspect (cornmeal, grits, etc.).

Aflatoxins are difficult to remove once produced. High heat treatments, dissolving out with alcohol or acetone, treatment with strong acids or alkalies, and inactivation with ammonia, formaldehyde, hydrogen peroxide, or other chemicals are some methods that have worked. Chemicals that kill the fungi will prevent aflatoxins from being produced. Research is being done to find crop varieties that are resistant to fungal attack. There are natural substances in soybeans and some peanuts that reduce attack. Grain and meal should be stored under conditions that do not favor fungal growth. Contaminated feed is sometimes blended with good grain to reduce the aflatoxin to a "safe" level and then fed to "resistant" animals. Since only damaged or stressed crops are affected, doesn't it make sense to build up soil fertility and the health of the plant?

will lower the capacity of these crops, hybrid crops in particular, to produce seeds with high-protein content" (*Organic Gardening and Farming*, May 1960). Cattle will often graze next to a concrete or gravel road because the additional lime from the road improves forage quality. They even prefer to eat "weeds" growing on fertile soil over grass on poor soil (W. A. Albrecht, *Soil Fertility and Animal Health*, 1958, p. 113)

Are hybrids really better? One of the supposedly great boons to agriculture from applied science is the "improvement" of plants and animals from crossing different stocks or lines to obtain hybrids. These new hybrid varieties often have vastly different characteristics than their parents—greater yields, disease resistance, shorter maturity time, etc. Plant and animal breeding have brought us such wonders as tomatoes that are "engineered" to be of uniform size and to be firm enough to be able to be picked and packaged by machines—but they taste like cardboard!

Hybrids result from applying the laws of genetics to obtain a desired combination of characteristics in an offspring. The physical and functional characteristics of all organisms are controlled by their genes. Different individuals of a species carry different traits in their set of genes (there are thousands of individual genes). As we all know from human heredity, the children may or may not inherit dad's nose or grandma's curly hair. Geneticists identify "desirable" traits in various varieties of wild and domesticated plants and animals, then breed a pure inbred line having a certain trait. They then cross parents from two inbred lines to obtain a hybrid which they hope will have a certain combination of traits. Sometimes two hybrid lines (single crosses) will be crossed to give a double cross. Now scientists calling themselves genetics engineers are trying to manipulate and change single genes to give desired characteristics.

Is all this "tinkering" with nature really good? Some amazing things can be done, and at first glance, it seems great to have corn that is resistant to southern corn leaf blight and cows that give more milk with higher butterfat. One obvious disadvantage of hybrids is that they do not "breed true:" that is, planting seed from hybrid plants does not produce all hybrids in the next generation, because some of the offspring revert back to their ancestral traits. So the farmer is "locked in" to buying hybrid seed every year. Crossbred cattle may have good characteristics, but in the next generation, many undesirable offspring are born (*Western Livestock Journal*, Oct. 1969, p. 25–26).

Another problem with hybrids is that they are too limited in their characteristics—they are "fine tuned." They are all alike for certain traits (within a given variety), there is no variation. While certain "desirable" traits are present (are they always really desirable?), others are therefore missing which would give the plants or animals more adaptibility to varying conditions—weather, nutrition, pests. Natural populations of plant and animal species have a variety of gene traits in the different individuals; while a few may be unfit, most will be well adapted for their environment. An excellent example of the vulnerability of uniform hybrids is the wildfire spread of southern corn leaf blight in 1970. Because of these things, scientists are becoming concerned about the "genetic vulnerability" or "narrowing genetic bases" of our crops through hybridization, and the possible loss of the original wild species of plants (from extinction by man's activities) from which the domesticated varieties were bred (Jack Harlan, *Journal of Environmental Quality*, 1973, vol. 1, no. 3, p. 212). "Gene banks" of stored seeds from many native varieties of crop plants are now being maintained in order to preserve genetic traits that may prove useful in the future (*Science*, 1981, vol. 214, p. 161–164, 421–423).

What are the "desirable" traits that agricultural geneticists seek for in hybrids? Too often they emphasize high yields—quantity, not quality. While there are certain strengths, there are also weaknesses. If you press a hybrid corn breeder, he will admit that the various varieties are "fine tuned" to perform well on different soils—soils with one or another deficiency or fertility imbalance, such as high magnesium or high potassium, and some will even say that they do not have a variety that will do well on "good" soil. Tests by Ernest M. Halbleib of McNabb, Illinois, have found that hybrid corn did not absorb 7 to 9 trace minerals. None of the varieties he tested took up cobalt, a mineral needed in vitamin B_{12} which animals need for good health. (*An Acres U. S. A. Primer*, 1979, p. 38). A cobalt deficiency has been linked to the development of brucellosis and undulant fever. According to the University of Nebraska Experiment Station Bulletin 144 (1946), open pollinated corn makes chicks gain faster and have more protein in their meat than hybrids, and it makes far superior quality silage.

The "green revolution" of the late 1960's seemed like an agricultural savior at first glance. The new hybrid varieties of corn, wheat, and rice, produced greatly increased yields—but at what cost? They cannot grow successfully unless given large doses of fertilizers and protected by pesticides, which underdeveloped countries cannot afford, plus the native people do not like their flavor. In spite of these problems, new varieties are still being developed and introduced into Third World countries (*Science*, vol. 217, July 16, 1982, p. 215–220).

Theoretically, not all hybrids are bad. It just depends on what particular combination of traits (genes) wind up in the hybrid variety. Do they make the plant or animal more vigorous and give more nutritious food, or do they produce an out-of-balance plant or animal that gives "foodless food"?

A growing number of farmers are finding that open pollinated crops can produce yields nearly equal to hybrids, but of much higher quality. Adolph Steinbroun of Fairbank, Iowa, had his open pollinated corn tested. Compared to average midwestern test figures, his corn had 75%

Poor and good root development in corn from the same field, but under different fertilizer programs.

more crude protein, 875% more copper, 345% more iron, and 205% more manganese (*An Acres U. S. A. Primer*, 1979, p. 38). It all traces back to the soil.

Quality—the role of fertile soil. We have seen in Chapter 16 that soil fertility—good or poor—can affect the resistance of plants to diseases and pests. This happens through internal functions and processes of the plant; that is, its metabolic functions are different. Is it any wonder then that soil fertility can also influence the food quality of crops—their protein, vitamin, and mineral content?

Even though some "experts" have claimed that soil fertility does not affect the vitamin or nutrient content of crops, numerous studies have proved that it does. Tests run on vegetables from different areas have found a four-fold phosphorus variation and a 32-fold iron variation in spinach. Tomatoes have been found to vary over two times (200%) in phosphorus content, nearly three times in potassium, five times in calcium, over six times in boron and cobalt, 12 times in magnesium, 53 times in copper, 68 times in manganese, and 1938 times in iron (J. I. Rodale, *The Complete Book of Composting*, 1972, p. 595–596). Dr. William D. McElroy of Johns Hopkins University found that unbalanced trace element nutrition can interfere with a plant's enzyme system and make it inferior in quality long before it appears stunted or shows leaf discoloration. He found that iron and copper deficiency reduced enzyme activity in tomatoes by over 80% (*The Complete Book of Composting*, p. 612). Dr. Charles Northern has found that adding a small amount of manganese increased the vitamin C content of plants three times, while boron increased vitamin A in apples 100% (*The Complete Book of Composting*, p. 614). Experiments growing peppermint in complete and deficient nutrient solutions showed that the amino acids produced by plants are unbalanced if the plant's nutrition is deficient (G. R. Noggle & G. J. Fritz, *Introductory Plant Physiology*, 1976, p. 251; see box).

PERCENT OF ALCOHOL-SOLUBLE AMINO ACIDS AND AMIDES IN LEAVES OF PEPPERMINT GROWN IN VARYING NUTRIENT SOLUTIONS.

amino acid or amide	*complete nutrients*	*potass. defic.*	*calcium defic.*	*phosph. defic.*	*sulfur defic.*
glutamine	12.7	50.1	6.0	18.7	9.2
leucine	1.8	6.7	1.8	1.2	-
valine	3.1	4.3	2.4	2.5	1.2
alanine	6.0	4.7	4.7	3.2	1.3
serine	3.6	3.6	2.1	3.3	1.2
S-amino butyric acid	6.7	2.8	5.6	3.6	5.6
glutamic acid	38.8	3.5	25.4	30.7	3.3
aspartic acid	27.3	6.6	10.3	26.0	3.3
asparagine	-	15.6	29.2	10.8	16.4
tyrosine	-	2.1	-	-	-
proline	-	-	9.7	-	-
arginine	-	-	-	-	58.5
methionine	-	-	2.8	-	-

High levels of potassium or low levels of phosphorus increase the nitrate content of plants, while balanced fertilizer levels produce highest quality silage (Research Report 29, Part 1, Univ. of Wisconsin College of Agriculture Experiment Station, Oct., 1967, p. 2, 7).

Besides the beneficial effects on the physical characteristics of the soil (tilth, aeration), organic matter (humus and the microorganisms living in it) has a great deal to do with producing high quality crops.

Calcium–Potassium

• From W. A. Albrecht, "Soil Fertility as a Pattern of Possible Deficiencies," *Journal of the American Academy of Applied Nutrition*, vol. 1, 1947, p. 17-28:

"The calcium-potassium ratio . . has given us a pattern of the protein possibilities in the crop. If nature, under less rainfall, has left much calcium in the soil, we have a proteinaceous crop. If the soil is under higher rainfall to give a small amount of calcium in relation to potash, then we have a carbonaceous crop. The validity of this belief, namely that a liberal supply of calcium in the soil in relation to potassium represents production of crops rich in proteins and minerals, while the reverse relation give crops high in carbohydrates—thereby low in proteins and minerals—was tested. Soybeans were grown with increasing amounts of potassium available in the soil and associated with constant amounts of calcium. Three ratios of calcium in potassium were used while all other nutrients were liberally supplied. Increasing the potassium increased the forage yield to a maximum of 25%. This fact would draw ready applause for an experimenting agronomist. Such work can win funds in support of it as research. But the buffalo of the western plains didn't evaluate herbage in terms of bulk. Our livestock does not use that criterion either. Hence, while increase of bulk may appear laudable, fixing our attention on bulk in relation to soil fertility has been leading us to grow more crops with serious deficiencies as feeds.

"Chemical analyses were made of the forage. The nitrogen content of the smallest of the three crops was 2.8%; of the intermediate crop 2.5%; and of the largest crop 2.19%. While we increased the bulk 25%, we reduced the concentration of nitrogen, and therefore the protein, by more

continued . . .

Growth-promoting substances are produced in the soil, trace elements are supplied to plants in more available forms, and crops contain more vitamins than those grown without organic matter (S. A. Waksman, *Proceedings of the Soil Science Society of America*, vol. 7, 1942, p. 20–21; *Humus*, 1938, p. 360).

There is no substitute for healthy, living soil. Without it, truly nutritious, high quality crops cannot be grown, and healthy, high-producing animals cannot be raised.

Why bother? O.K., you say, so high quality crops are better. But the market pays for bushels and pounds. What's the use of growing high quality crops if I can't sell them for a good price?

Of course the value of growing high quality crops is obvious if you are going to feed them to your own livestock or eat them yourself. The problem comes when you raise crops to sell. Unfortunately, the general

Calcium–Potassium, continued

than that figure. So that the greater amount of total protein was not in the largest but in the smallest crop.

"The phosphorus concentration, by analysis, was .25% in the small crop; .18% in the intermediate; and .14% in the largest crop. Assuming that the cow could digest it completely, she would be compelled to eat approximately twice as much of the larger crop to get the same amount of phosphorus. In the case of the calcium, this was approximately .75% of the dry weight in the smallest crop and only .27% or about one-third as much in the largest crop. Can any cow increase her consuming capacity by three times? We can't expect her to become a hay baler.

"We need to be concerned not only with the bulk of the crop, but also with the synthetic operations of the plant in using the fertility elements from the soil to convert the carbonaceousness over into proteinaceousness. Those processes make "grow" foods instead of merely "go" foods. They must be more generally appreciated if we are not to invite nutritional deficiencies more commonly."

• From Dr. Frank A. Gilbert, in his book, *Mineral Nutrition of Plants and Animals* (1948), p. 43, 45:

"The present opinion seems to be that a high potassium content of the soil when not balanced by other essential elements, especially calcium, tends to produce a plant especially high in carbohydrates. A normal amount when calcium and phosphorus are deficient will have the same effect and will produce plants of a very low biological value with a lesser change in caloric value. On the other hand, a more nutritious proteinaceous and less carbonaceous plant grows when the potassium-calcium ratio is weighted in favor of the latter element."

market is mainly only interested in bulk, in quantity, not quality. So where can a quality-conscious farmer find markets?

Markets. First of all, some markets do pay a premium price for high quality, such as for Grade A milk or high test weight grain, but you should be able to get better returns by seeking non-conventional markets. There are a surprising number and variety, including:

1. Organic food outlets and CO-OPs. If you farm entirely without synthetic chemicals, you may be able to become certified as an organic grower. Each state or region has its organic marketing network; there are about 35 regional groups that are active in 29 states.

2. Supermarkets, bakeries, restaurants, etc. Certain food wholesalers and retailers are interested in high quality products (and others are not). Ask in your local area. Some breakfast cereal manufacturers want quality grain.

3. Foreign markets. Most foreign grain importers definitely want high quality grain and will pay premium prices. You will need to make contacts to deal directly with the foreign buyers and bypass domestic markets.

4. Bird food. Poultry raisers, egg farmers, and bird seed manufacturers will pay high prices for quality grain.

5. Vegetables and fruits. Raising a few acres of truck crops or starting an orchard can pay large dividends, not only through numbers 1 and 2, above, but also through roadside fruit and vegetable stands and farmers markets. Retail sales bring larger profits than wholesale.

6. Check the advertisements in the publications that promote organic and ecologically sound agriculture, including *Acres U.S.A.*, *The New Farm*, *Natural Food and Farming*, *Organic Gardening*—or run your own ad in these or in local farm papers or "shoppers."

Some organizations that can be contacted for additional help are:

1. Natural Food Associates
   P.O. Box 210, Atlanta, TX 75551

   A national, non-profit organization with chapters in many states. Publishes *Natural Food and Farming*.

2. International Federation of Organic Agriculture Movements
   Les Quatres Saisons, 6 rue Saulnier, 75009 Paris, France

   Coordinates organic farming developments in 30 nations. Publishes a quarterly bulletin in four languages.

3. Regenerative Agriculture Association
   222 Main Street
   Emmaus, PA 18049

   Promotes and does research on self-renewing farming. Emphasis on bio-technology, elimination of chronic joblessness, and renewing the people-land link. Publishes *The New Farm*.

4. National Farmers Organization
   Corning, IA 50841

   Promotes collective bargaining and parity for agriculture; runs marketing facilities. Publishes *The NFO Reporter*.

5. Acres U.S.A.
   P.O. Box 9547, Kansas City, MO 64133

   Promotes "eco-agriculture." Publishes *Acres U.S.A.*

6. National Organization for Raw Materials (NORM)
16039 Prairie Villa St. S.W., Tenino, WA 98589

Provides a forum to educate Americans in the essential rolė of raw material economics in national prosperity. Publishes a *Progress Report.*

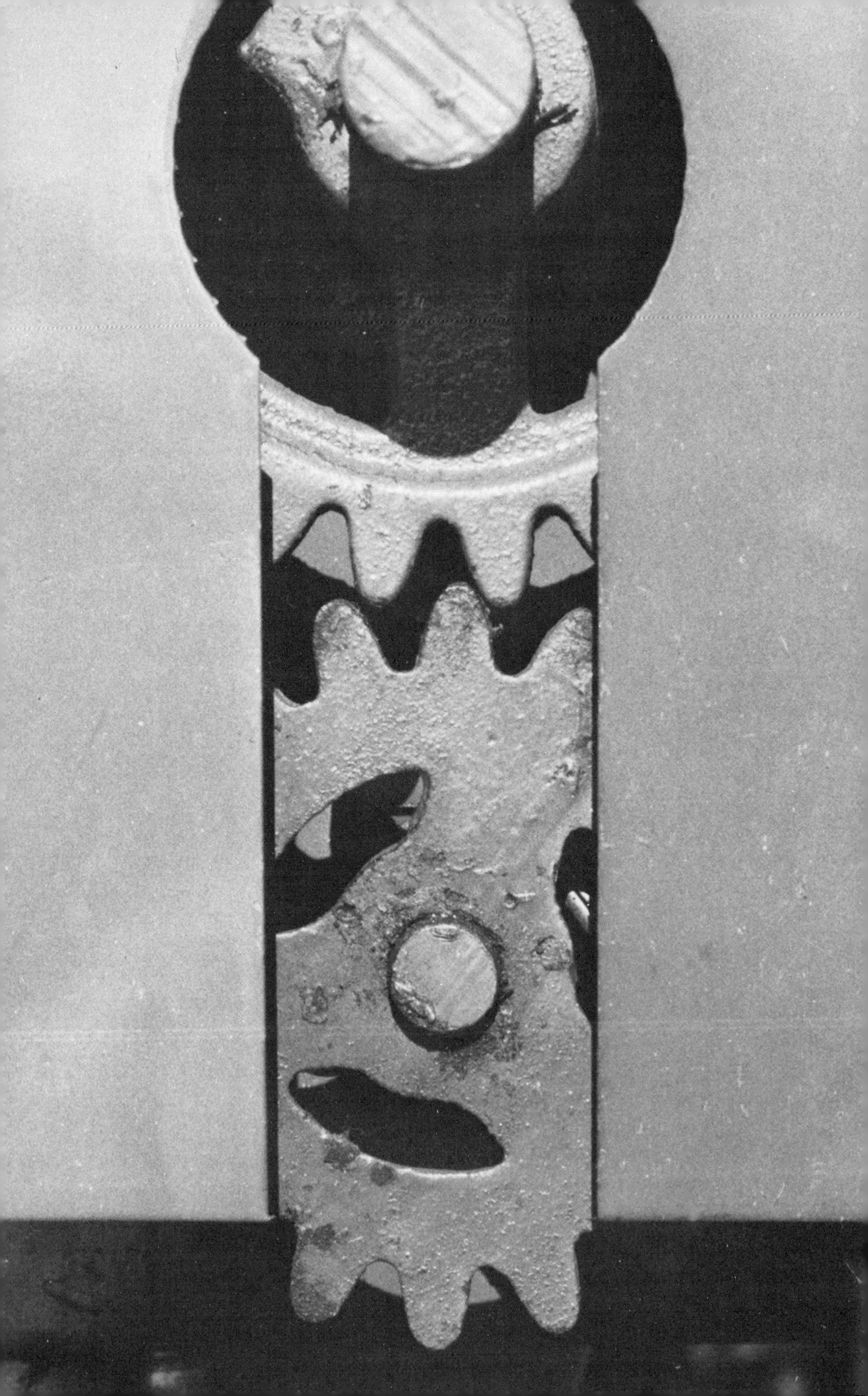

CHAPTER 18

A Cog in the Wheel?

If you have read this far, you are obviously not an average person—*you care*. You are concerned about problems and want to find solutions. You are not just another cog in the wheel of modern agriculture. *You want to think for yourself.*

We have taken a journey searching for "the rest of the story" about agriculture today. We have taken many paths and detours. We have looked at nature's systems and man's system. We have seen how a plant makes food and how a manufacturer makes a fertilizer formula. We have found some interesting and surprising things.

I hope you have become convinced that for the most part, "modern" agriculture is misguided, heading in the wrong direction. *There is a better way!* I hope you are convinced that "natural-systems" agriculture is the way to go. It is the wave of the future. It is ahead of its time. It is actually more scientific than what the "experts" call scientific agriculture, for it is based on natural laws and systems, and seeks to work *with* them, not against them.

Natural-systems farming is not impractical—*it works*! Once you get your soil turned around and into a healthy state, *things will really happen*: high yields of crops with very high nutritional value—quality—and healthy, high-producing animals. And your family's health will also improve. And probably at no more expense that what you spend now for fertilizers, herbicides and pesticides, and vet and doctor bills. Of course there will be a transition period while things are turning around, while the poisons are being cleaned out of your soil and the beneficial organisms are being reestablished. That is a delicate time when you need patience and optimism, and plenty of reliable advice; do not hesitate to pay for good consultants. But you will see encouraging results right from the beginning, and crops will improve steadily year after year.

It takes guts. It takes courage to break out of the mold. To be different from your neighbors. To try something new. To change.

Agriculture has to change. The present system is leading us downward in an ever-increasing spiral. It takes concerned and far-sighted individuals to break out of the system and go on to better things. *You can do it*—if you really want to!

Glossary

aerobic–with oxygen present (see anaerobic)
anaerobic–without oxygen (see aerobic)
anion–a negatively-charged ion (see cation)
atom–the smallest particle of matter (an element) which has the properties of that element
cation–a positively-charged ion (see anion)
cell–the smallest unit of living organisms
chelate–a water soluble organic compound that can hold a metal cation
chemical–any particular kind of matter (see atom, compound, element, ion, molecule)
colloid, colloidal–a material of very small particle size (less than 0.002 mm, or 1/5000 inch), but larger than most molecules; has a large surface area
compound–a material composed of two or more elements
ecology–the study of organisms and their environment, and the interrelationships among them
ecosystem–a total ecological system composed of all of the living organisms plus the non-living environment
element–the simplest type of matter, composed of only one type of atom
enzyme–a substance which increases the speed of metabolic or other chemical changes, usually inside living cells
foliar–referring to leaves
gene–the smallest unit of heredity; found on chromosomes in every cell; determines the inherited characteristics of living organisms
hormone–a substance secreted by a plant or animal that regulates developmental processes: metabolism, growth, and reproduction
humus–a colloidal material resulting from decomposition of plant or animal organic matter
inorganic–chemical definition: chemical compound not containing carbon; ordinary definition; not produced by living organisms (see organic)

ion–an electrically charged atom or group of atoms

metabolism, metabolic–any of the chemical activities of living cells or organisms

microbe–same as microorganism

microorganism–a microscopic organism

molecule–the smallest particle of a compound, composed of two or more atoms or elements

organic–chemical definition: chemical compound containing carbon; ordinary definition: referring to or produced by living organisms (see inorganic)

organism–any living thing, whether microbe, plant, or animal

osmosis–seeping (diffusion) of water through a thin membrane, from a higher concentration to a lower concentration of water

parasite–an organism that lives on or in the body of another organism and derives its nutrition from the other organism

pathogen–a disease-causing microorganism

pH–a measure of acidity or alkalinity, with pH 7 being neutral, pH 0 to 7 acid, and pH 7 to 14 alkaline

photosynthesis–producing food (sugar) by green plants, using sunlight

physiology, physiological–the functioning of cells or organisms

predator–an organism (animal) that gets its food by killing and eating other animals

respiration–releasing energy from food by breaking it down inside the organism's cells

salt–a chemical compound formed from ions from an acid and a base

species–a certain kind of organism

stoma, stomata, stomates–tiny openings on the surface of leaves that allow gas exchange

tilth–a crumbly soil texture

tissue–a group of cells having a particular function

translocation–transfer of food or other products from one part of a plant to another

APPENDIX

Basic Chemistry

Although we cannot present an entire chemistry course here, we will give enough of the principles and terminology of chemistry to aid in understanding soils, fertilizers, and plant functions. First some terms and concepts.

1. Chemical. Not a "bad word," but merely the name for any kind of matter, whether gas, liquid, or solid. All matter, including yourself, is composed of chemicals.

2. Organic vs. inorganic. "Organic" is sometimes used by farmers, gardeners, and health enthusiasts as synonymous with "natural," or produced by living organisms (this was the original meaning). The chemical definition of organic is carbon-containing compounds; all other compounds are called inorganic (most inorganic chemicals are "natural," such as oxygen, nitrogen, and sulfur).

3. Compound. A pure substance that can be decomposed into two or more different elements. Examples of compounds include water, carbon dioxide, and sugar.

4. Element. A type of matter that cannot be decomposed into simpler substances. There are only 90 elements that occur in nature (plus others made by man). Each element is given a name and an abbreviation, or chemical symbol, such as H for hydrogen, O for oxygen, etc. (see box).

5. Atom. Smallest particle of an element that can enter into chemical changes (but composed of subparticles: protons, neutrons, and electrons, forming a "miniature solar system").

6. Molecule. Smallest particle of a compound, or a stable combination of atoms of one element, such as O_2, Cl_2, H_2O, CO_2, H_2CO_3.

7. Ion. An atom or group of atoms carrying an electrical charge. There are two kinds, cations (positively charged, such as H^+, Na^+, Ca^{++}) and anions (negatively charged, such as Cl^-, O^{--}, SO_4^{--}). The number of + or − charges determines how many ions

are needed to combine to form a molecule, such as H_2O, $CaCl_2$, CaO.

8. Bonds. Atoms and ions are held together by various kinds of chemical bonds, which are an attractive force, a type of stored energy, which is released when the bonds are broken.

Reactions and equations. Many chemical changes (reactions) involve compounds that act as acids and bases. Typical acids contain hydrogen ions (H^+); typical bases contain hydroxyl ions (OH^-). Acids react with bases to form salts and water. We can write the reaction using chemical formulas; the reaction is called an equation, and the number of atoms of each element must be balanced on each side of the arrow. For example:

hydrochloric acid + sodium hydroxide yields sodium chloride + water

$$HCl + NaOH \rightarrow NaCl + H_2O$$

Another reaction, involving sulfuric acid is:

sulfuric acid & sodium hydroxide yields sodium sulfate + water

$$H_2SO_4 + 2\,NaOH \rightarrow Na_2SO_4 + 2\,H_2O$$

Other types of chemical reactions are similarly written:

1. Two gases combine: hydrogen + oxygen yields water
   $2H_2 + O_2 \rightarrow 2H_2O$

2. Decomposition: ammonium chloride yields (when heated) ammonia + hydrogen chloride
   $NH_4Cl \rightarrow NH_3 + HCl$

3. Simple replacement: iron + sulfuric acid yields ferrous sulfate + hydrogen
   $Fe + H_2SO_4 \rightarrow FeSO_4 + H_2$

4. Double decomposition: calcium chloride + sulfuric acid yields calcium sulfate + hydrogen chloride
   $CaCl_2 + H_2SO_4 \rightarrow CaSO_4 + 2\,HCl$

CHEMICAL ELEMENTS ESSENTIAL TO PLANTS

element	*chemical symbol*	*forms available to plants*
carbon	C	CO_2 (carbon dioxide)
oxygen	O	O_2 (oxygen gas), H_2O (water)
hydrogen	H	H_2O (water)
nitrogen	N	NO_3^- (nitrate), NH_4^+ (ammonium)
calcium	Ca	Ca^{++} (calcium ion)
phosphorus	P	$H_2PO_4^-$ (dihydrogen phosphate, orthophosphate), HPO_4^{--} (hydrogen phosphate)
magnesium	Mg	Mg^{++} (magnesium ion)
potassium	K	K^+ (potassium ion)
sulfur	S	SO_4^{--} (sulfate)
chlorine	Cl	Cl^- (chloride)
iron	Fe	Fe^{+++} (ferric), Fe^{++} (ferrous)
copper	Cu	Cu^{++} (cupric), Cu^+ (cuprous)
zinc	Zn	Zn^{++} (zinc ion)
boron	B	BO_3^{--} (borate), $B_4O_7^{--}$ (perborate)
manganese	Mn	Mn^{++} (manganous)
molybdenum	Mo	MoO_4^{--} (molybdate)

Helpful Reading

Albrecht, William A. 1975. *The Albrecht Papers* (edited by Charles Walters, Jr.). Acres U.S.A., Raytown, MO. 515 p.

Campbell, Stu. 1975. *Let it Rot!* Garden Way Publ., Charlotte, VT. 152 p.

Harmer, Ruth M. 1971. *Unfit for Human Consumption.* Prentice-Hall, Inc., Englewood Cliffs, NJ. 374 p.

Longgood, William. 1960. *The Poisons in Your Food.* Simon & Schuster, NY. 277 p.

McVickar, Malcomb H. 1970. *Using Commercial Fertilizers*, 3rd ed. The Interstate Printers & Publishers, Danville, IL. 352 p.

Organic Gardening & Farming Staff. 1966. *The Organic Way to Plant Protection.* Rodale Press, Inc. Emmaus, PA. 355 p.

Tisdale, Samuel L. & Werner L. Nelson. 1974. *Soil Fertility and Fertilizers*, 3rd ed. The Macmillan Co., NY. 694 p.

Van Gorder, Dan P. 1966. *Ill Fares the Land.* Western Islands Publ., Boston, MA. 240 p.

Walters, Charles J. 1971. *Unforgiven.* Economics Library, Kansas City, MO. 458 p.

Walters, Charles Jr. 1975. *The Case for Eco-Agriculture.* Acres U.S.A., Raytown, MO. 122 p.

Walters, Charles Jr. & C. J. Fenzau. 1979. *An Acres U.S.A. Primer.* Acres U.S.A., Raytown, MO. 465 p.

Whiteside, Thomas. 1971. *The Withering Rain; America's Herbicide Folly.* E. P. Dutton & Co., Inc., NY. 224 p.

INDEX

(Chemical symbols and formulas are indexed after their name)

A

F

G

H

I

K

L

M

N

O

P

Q

R

S

T

U

V

W

X

Y

Z